The Biological Rhythms and Clocks of Intertidal Animals

The Biological Rhythms and Clocks of Intertidal Animals

JOHN D. PALMER

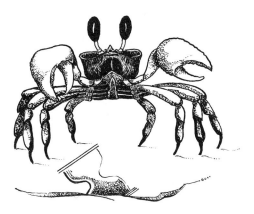

New York   Oxford
OXFORD UNIVERSITY PRESS
1995

Oxford University Press

Oxford   New York   Toronto
Delhi   Bombay   Calcutta   Madras   Karachi
Kuala Lumpur   Singapore   Hong Kong   Tokyo
Nairobi   Dar es Salaam   Cape Town
Melbourne   Auckland   Madrid

and associated companies in
Berlin   Ibadan

Copyright © 1995 by Oxford University Press, Inc.

Published by Oxford University Press, Inc.,
200 Madison Avenue, New York, New York 10016

Oxford is a registered trademark of Oxford University Press

Library of Congress Cataloging-in-Publication Data
Palmer, John D., 1932–
The biological rhythms and clocks of intertidal animals/
John D. Palmer.
p. cm.   Includes bibliographical references and index.
ISBN 0-19-509435-2
1. Intertidal fauna. 2. Biological rhythms. I. Title.
QL121.P343   1995
574.1′882′09146—dc20   94-25802

2 4 6 8 9 7 5 3 1

Printed in the United States of America
on acid-free paper

This book is dedicated to:

a talented colleague, Barbara Williams, who manages to retain her wonderful sense of humor even while struggling to work with me . . .

and to the memory of Mike Laverack who, along with his wife, was recently killed in a helicopter crash — he was a good friend, a fine scientist, a polymath, a witty epistolist, and one of those rare individuals who actually answered letters written to him.

Preface

This book is essentially a sequel to my 1974 monograph, *Biological Clocks in Marine Organisms*. In the intervening 20 years, a great deal of fine new research has been carried out on this subject, and also on that of circadian rhythms. The former is the substance of this monograph, and the latter is integrated as needed — especially in the last chapter. The result is a disquisition about the progress in our understanding of the still undeciphered clockworks of the living horologes that forge animal behavior and physiology into waveform outputs, which the romantics among us call "biological rhythms." The last chapter discusses in detail the more recent "circalunidian clock hypothesis."

It is because living clocks cause *oscillatory* homeostasis to be the rule (supplanting Claude Bernard's earlier straight-line stasis), and because the clocks are distributed widely throughout all eukaryotic life, that it is imperative that the clockworks be deciphered — the sooner the better. The importance of the project should certainly result in a Nobel Prize for the discoverer.

Biological rhythms come in several versions, each defined by its fundamental period length. The main versions studied are daily, tidal, and year-long rhythms. Of the three, most work has been carried out on daily rhythms, and the least on yearly ones ("publish or perish" impacts heavily on the latter; the study-time required means papers can be produced only infrequently). For two reasons the most difficult to research are tidal rhythms. First, the clocks governing them are maddeningly noisy. Second, the nature of the tides is complicated: they have ever changing period lengths, meaning that without some special understanding of the cosmic generation of tides workers cannot even know what output to expect from an intertidal-dwelling research subject. Those two stigmas drive people, otherwise interested in chronobiology, to work on daily rhythms whose period simply matches that of the steady rotation of the earth. To overcome those two hurdles immediately, this book begins with a description of how tides are generated, and why organismic tide-associated rhythms are so noisy (Chapter 1). Next, methods for overcoming — or at least tolerating — the noise are discussed (Chapter 2), and then the rest of the book focuses on rhythms, properties thereof, and the nature of the clocks governing them.

I have attempted to make the book understandable to anyone who has completed successfully a respectable college-level introductory biology course; but obviously, the more biology one has taken, the easier the reading will be. The intended audience is upper-level undergraduate students, graduate students, behaviorists, physiologists, marine biologists, and my colleagues in chronobiology. The book should be of particular interest to those who work on daily rhythms: (1) it will bring them up to date on the progress of tidal-rhythm research; and (2) it will give them a feeling as to how difficult (relative to their branch) working with tide-associated rhythms can be. In a nutshell, the study of tide-related organismic rhythms is a playground for people who like to rise to a challenge, and limbo for fanatics who do not know when to quit.

Over the years, chronobiologists have adopted a bad habit that some may not even be aware of: in their publications many authors tend to hold conversations only with other club members. Not only have they developed their own argot — much of which is unintelligible even to other biologists — they even "talk" in symbols. I need give only one example, which I guarantee is taken verbatim from the literature, from which everyone will understand the point of my concern:

> It is clear . . . that as $-\Delta\phi_E$ becomes larger, in the early subjective night, so does the slope of the PRC: it follows that as τ is shortened and the $-\Delta\phi_E$ necessary for entrainment increases, the evening light pulse falls on the PRC where the S_E is steeper and seasonal variation in ϕ_E and ϕ_E' are accordingly further reduced.

The above flaunts nicely an author's Greek ancestry, and can be a fine means of communication . . . but only for those conversant in that idioglossia. Few are however, and none should be subjected to such a challenge to comprehension. In fact, sophisticated readers usually ascribe the use of such technogarble to an attempt to add, via the pen, a profundity absent from the content of the paper. It makes us suspect that the clothes have no emperor. Be assured that no claptrap will appear in this book, nor will highly specialized chronobiological terminology. Additionally, a glossary of terms appears at the end of the book.

In writing a monograph, a fine library must be at one's fingertips.

Thanks to a past Governor of Massachusetts being more interested in running for President than running the State, nearly half of the biological-journal subscriptions in my University of Massachusetts library had to be cancelled. This book would not — could not — exist if it were not for the superb biology library maintained by the Marine Biological Laboratory in Woods Hole, Massachusetts. It is clearly one of the finest in the world (philanthropists take note; journal subscriptions are not immune from worldwide inflation). And a crutch, not available to me in previous book endeavors, is the gift of The Internet: Using e-mail, virtually any questions I had were quickly answered if an internaut correspondent remembered to log on to cyberspace occasionally. What a wonderfully efficient way to communicate. (I can be reached at ftodd@bio.umass.edu)

Marine Biological Lab   J.D.P.
Woods Hole, Massachusetts
October 1993

Figure Credits

The figures listed below were reproduced with the permission of the following sources using this format: Publisher (journal name), chapter number: figure numbers.

Academic Press (Palmer, *An Introduction to Biological Rhythms*), Chapter 5, Fig. 5.

Academic Press Ltd. (*Animal Behavior*), Chapter 4: 33.

Academic Press Ltd. (*Journal of Theoretical Biology*), Chapter 2: 6.

Academic Press Ltd. (Sleigh and Alister, *The Effects of Pressure on Organisms*), Chapter 4: 2.

American Institutes of Biological Sciences (*BioScience*), Chapter 2: 7; Chapter 3: 9.

American Museum of Natural History (*Natural History Magazine*), Chapter 5: 2.

American Society of Zoologists (*American Zoologist*), Chapter 5: 3, 4.

Cambridge University Press (*Journal of the Marine Biological Association of the U.K.*), Chapter 3: 29; Chapter 4: 3, 15, 19–21, 30, 31; Chapter 5: 12.

Center for Agricultural Publishing, PUDOC (Bierhuizen, *Circadian Rhythmicity*), Chapter 6: 9.

Company of Biologists (*Journal of Experimental Biology*), Chapter 3: 11–13; Chapter 6: 17, 18.

Elsevier Science Publishers (*Journal of Experimental Marine Biology and Ecology*), Chapter 3: 44; Chapters 4: 4–7, 33, 34.

Elsevier Science Publishers (*Journal of Insect Physiology*), Chapter 5: 13.

Elsevier Science Publishers (Naylor and Hartnoll, *Cyclic Phenomena in Marine Plants and Animals*), Chapter 5: 17.

Gordon & Breach Science Publishers (*Marine Behaviour and Physiology*), Chapter 2: 2–5, 8; Chapter 3: 2–8, 10, 15–28, 30; Chapter 4: 10–13, 22–25, 28, 29; Chapter 6: 3–6.

Inter-Research (*Marine Ecological Progress Series*), Chapter 4: 8, 9.

JAPAGA (Aldrich, *Quantified Phenotypic Responses in Morphology and Physiology*), Chapter 3: 31–35.

John Wiley & Sons (*Journal of Cell and Comparative Physiology*), Chapter 6: 14.

John Wiley & Sons (Palmer, *Biological Clocks in Marine Organisms*), Chapter 1: 1–5; Chapter 3: 43; Chapter 5: 1.

John Wiley & Sons (Rosenberg and Runcon, *Growth Rhythms and the History of the Earth*), Chapter 3: 36, 37.

S. Karger Publishers (Pevet, *Comparative Physiology of Environmental Adaptations*), Chapter 5: 16.

Macmillan Magazines (*Nature*), Chapter 3: 38; Chapter 6: 11–13.

Marine Biological Association (*Biological Bulletin*), Chapter 2: 1; Chapter 3: 41, 42; Chapter 5: 7, 9, 18.

Oxford University Press (*Journal of Neuroscience*), Chapter 6: 8.

Raven Press (*Chronobiology International*), Chapter 1: 6; Chapter 6: 16.

Springer-Verlag (*Journal of Comparative Physiology*), Chapter 4: 16, 18; Chapter 5: 10, 11; Chapter 6: 1, 2.

Springer-Verlag (*Oecologia*), Chapter 5: 8.

Wayne State University Press (*Human Biology*), Chapter 3: 39.

The figures listed below were purchased from the sources indicated. The same format used above is repeated here.

Harvard University Press (Moore-Ede, Sulzman, and Fuller, *The Clocks That Time Us*), Chapter 4: 14.

Scientific American, Inc. (*Scientific American*), Chapter 3: 1, 40; Chapter 6: 15.

University of South Carolina Press (DeCoursey, *Biological Rhythms in the Marine Environment*), Chapter 5: 14, 15.

Special thanks to Margaret Nutting, the artist who made the nine vignettes for the book. The drawings are entitled (in their order of appearance): Ghost crab welcome (title page); Fiddler crab setting its clock; Crab nerd at PC; Dancing to rhythm of the tides; Schizo crab bifurcation; Intertidal virtuoso; Palolo's burlesque of spring tides; Interphyletic comparison of clock types; Ying/Yang of organismic rhythms.

J.D.P.

Contents

1. **Introduction to Organismic Rhythms and Tidal Cycles**   3
   The Tides   5
   Literature Cited   13

2. **Time-Series Analysis**   14
   Tidal Rhythm Model   17
   Subtle Dual Period Model   19
   Combined Tidal and Daily Rhythm Model   21
   Tidal Rhythm Combined with an Ultradian Cycle   23
   Peaks That Scan the Day at Different Rates   23
   Conclusions on Models   24
   Array Analysis and Compact Plots   26
   Literature Cited   29

3. **A Survey of Tide-Associated Rhythms**   32
   Fiddler Crabs (genus *Uca*)   32
   The Green Shore Crab (*Carcinus maenas*)   43
   The Penultimate-hour Crab (*Sesarma reticulatum*)   50
   The Cranny Crab (*Cyclograpsus lavauxi*)   54
   The Pliant-Pendulum (*Helice crassa*) and Schizo
      (*Macrophthalmus hirtipes*) Crabs   56
   Peracarids   59
   The New Zealand Clockle (*Austrovenus stutchburyi*)   63
   The Basket Cockle (*Clinocardium nuttalli*) as a
      Geochronometer   68
   Fishes   69
   Are There Moon-Related Rhythms in Humans?   73

Miscellaneous Cases   75
Literature Cited   83

4. Phase Setting and Entrainment   89
The Green Shore Crab (*Carcinus*)   89
The Estuarine Crab (*Rhithropanopeus harrissii*)   105
The Portunid Crab (*Liocarcinus holsatus*)   105
The Rocky Shore Crab (*Hemigrapsus*) and its Phase-Response
    Curve   106
The Isopod (*Excirolana chiltoni*)   109
The Isopod (*Eurydice pulchra*)   112
The Amphipod (*Corophium volutator*)   114
The Amphipod (*Synchelidium sp.*)   120
The Amphipod (*Bathyporeia pelagica*)   120
Mollusks   120
The Shanny (*Lipophrys pholis*)   121
Literature Cited   132

5. Persistent Fortnightly and Monthly Rhythms   135
The Polychaete (*Typosyllis prolifera*)   138
The Flatworm (*Convoluta roscofensis*)   140
Another Flatworm (*Dugesia dorotocephala*)   141
The Land Crab (*Sesarma haematocheir*)   143
The Land Crab (*Cardisoma guanhumi*)   148
The Fiddler Crab (*Uca*)   149
The Amphipod (*Talitrus saltator*)   150
The Isopod (*Eurydice pulchra*)   150
Bees   152
The Ant Lion (*Myrmeleon obscurus*)   152
An Intertidal Midge (*Clunio marinus*)   153
Epilogue   156
Literature Cited   157

6. The Elucidation of the Clock   161
Attacking the Clock and Coupler with Chemicals   162
Genetic Search for the Clock in the Fruit Fly   170
Anatomical Location of Clocks   172
The Role of the Eyestalks   183
The Clock Hypotheses   192
Literature Cited   196

Glossary   205
Author Index   209
Subject Index   213

The Biological Rhythms and Clocks
of Intertidal Animals

1

Introduction to Organismic Rhythms and Tidal Cycles

Residing within most organisms is a pacemaker entity usually referred to by its *nom de guerre*, a **biological clock**. Actually, each individual contains more than a single clock: one, or more, is present, possibly, in every cell of multicellular plants and animals. This conclusion arises from studies in which parts of large organisms are isolated and kept alive in culture; there the tissues, though removed, continue to undergo typical clock-controlled **rhythms** in physiology. Further down the scale of living things, *Acetabularia*, a giant single-celled alga, has been demonstrated to contain multiple copies of the same clock in its cytoplasm!

Living clocks drive and mold an organism's physiology and behavior into cyclic displays that mimic the major geophysical periods on earth: the **solar day** (the 24 h interval between successive sunrises), the **lunar day** (the 24.8 h average interval between successive moonrises), the hemilunar day (the 12.4 h average interval between consecutive identical phases of the tides), the fortnight (the 14.77 day interval between, say, the peaks of spring tides (defined later in this chapter)), and the year (365.25 days).

The clocks that drive these particular rhythms have some special properties that other cellular pacemakers lack: (1) while obvious environmental cycles (such as day/night) can set the phase of some clocks (often referred to as setting the "hands of the clock") they do not provide the basic timing information to it; and (2) the rates at which clocks are virtually temperature independent, and are also immune to many chemical changes that organisms may be exposed to in the environment.

That these overt rhythms displayed by organisms are not simply driven by environmental cycles of the same frequency is simple to demonstrate. A rhythmic plant or animal is brought into the laboratory and placed in isolation where there are no day/night, tidal, or seasonal alternations. The lights are just left on or off constantly, and the temperature is unchanging. In this eternal uniformity, organismic cycles are found almost always to persist. Thus, it is surmised that each subject has within its body a living horologue. Experimental evidence has shown that the clocks are not built directly into a particular physiological process — such as the chain of chemical reactions of photosynthesis or respiration — causing it to be rhythmic. Instead, the clock is a separate unit with its own identity, which is coupled somehow to the process it causes to be rhythmic. This coupling sometimes breaks spontaneously, or can be forced to do so by manipulation in the laboratory. At the time of this writing, the clockworks and the means of coupling remain beyond our ken, although several generations of white coats have put an enormous amount of work into trying to decipher their mechanisms. The task has been comparable in difficulty to pushing a beached whale back into the sea at low tide.

While biological rhythms persist in the constancy of the laboratory, they usually change somewhat in period length, becoming slightly longer or shorter than the period expressed in nature. Most investigators are fond of saying that only in constant conditions can one see the natural **period** of a clock, because outside of the laboratory organismic rhythms are entrained (i.e., are synchronized) by and to ambient environmental cycles, such as the day/night alternation. This is misleading, because the rate at which solar-day clocks run in the constancy of the laboratory is a regular function of the intensity of the artificial constant light provided. This property is called Aschoff's Rule, named after a fine scientist who has contributed greatly to what is known about solar-day rhythms. This laboratory-engendered change in period is a bona fide property of all of these rhythms and was thus dignified by the prefix *circa* meaning "about". A **circadian rhythm** is thus a **solar-day rhythm** that in the laboratory has become *about a day* in length. Other neologistic conjunctions are circatidal, circalunidian, and the awkward circannual.

It should be obvious that the only time *circa* should be used, is when the lengthening or shortening property is displayed in the constancy of the laboratory. In nature the period is not *about*, it is *exactly* that of the environmental cycle to which the organism is exposed. However, verbal sloppiness over the years has allowed circadian to become synonymous with (solar) daily rhythm. But so be it, that usage is now thoroughly ingrained into chronobiological shoptalk. Too bad; after all we are scientists and as such are required to be precise. Therefore, in this book, although it may seem like "hypercorrect" niggling to some, whenever *circa* is used it will refer only to those rhythms studied in the laboratory whose period has been thus altered.

The references and a plethora of further information on the above outline of biological rhythms can be found in general works such as: Bünning (1973), Pengelley (1975), Palmer (1976), Moore-Ede *et al.* (1982), DeCoursey (1983), Gwinner (1986), Sweeney (1987), Edmunds (1988), Binkley (1990), and Hastings *et al.* (1991) — just to name a few.

So much for general background about organismic rhythms and clocks; the main thrust of this book is about the tide-associated rhythms of marine animals. To set the stage, I will start with a clear example, the activity rhythm of a fiddler crab (*Uca pugnax*). This is a semi-terrestrial crustacean that lives in the intertidal zone (that expanse of the shoreline between high and low tide) in self-made burrows during high tide inundations, and emerges out onto the mud flats during low tides to carry out a behavior usually described as the four Fs: feeding, fighting, fossicking and mating. A subject is brought into the laboratory, placed in an actograph (see Fig. 3-1), and exposed to atidal, constant conditions in an incubator. Here they are sheltered from all the major time cues provided by the environment. Their activity is measured automatically, without interrupting the constancy of the incubator, and passed to a data logger. A very fine example of one crab's display is seen in Fig. 1-1. On the first day of isolation, each time the tide receded on the old home shore, the animal ran in its actograph. During high tides the animal sat quietly in the incubator. The same pattern was repeated on subsequent days, but the period became slightly longer, i.e., as promised above, a *circa* period was assumed. In this example the period extended approximately 10 min to 13 h (Palmer, 1973).

The Tides

Next I must define several aspects of the tides themselves: how they are generated, and how and why they vary in the many ways they do. On completion of this exposé, the reader will have no choice but to come away with the question, why would Mother Nature have even tried to develop a living clock to be used to time an organism's behavior and physiology to such a messy, **discombobulated**, erratic cycle as the ebb and flow of the tides?

On most coastlines of the world there are two high and two low tides each day. The "day" referred to is not the solar day, it is the **lunar day**; the 24 h and 51 min average interval between successive moonrises. The tides are produced by gravitational and centrifugal forces generated

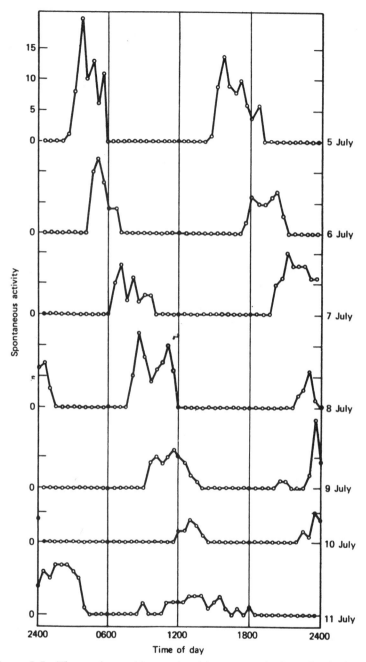

Figure 1-1 The persistent tide-associated locomotor rhythm of a single
fiddler crab (*Uca pugnax*) isolated in constant darkness, at a temperature of
20°C, and *sans* tides. The ordinate values are the number of tips of an
actograph/half hour. The peaks on the first day were in approximate synchrony
with the tides on the crab's home mudflat, but after that the period increased
to approximately 13 h (Palmer, 1974).

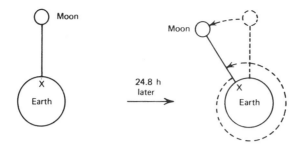

Figure 1-2   The relative movements of the moon and earth that produce the lunar day. The earth rotates on its axis in a counterclockwise direction (to an observer looking down on the north pole), while the moon orbits the earth in the same direction. After the earth has completed one rotation relative to the sun (a 24-h interval), the moon has traveled further along in its orbit. Thus, for the meridian signified by the X to be directly under the moon again, the earth must catch up by rotating another 12.75°. As a result, the interval between consecutive moonrises at a given longitude is (on average) 24 h and 51 min — the length of a lunar day (Palmer, 1974).

primarily between the moon and the earth, with some refinements added by the sun. The earth rotates on its axis relative to the sun once every 24 h, but it rotates once every 24.84 h relative to the moon. The additional 51 min in the latter case are the result of the fact that the moon, unlike the sun, is not a stationary reference point; instead, it is a body orbiting around the earth. Thus, as shown in Fig. 1-2, a particular longitude that starts with the moon directly overhead, must rotate 360° plus another 12.75° to "catch up" and have the moon again directly overhead. As the moon transits above the earth, its gravitational and tractive forces pull the sea below toward it making a broad, high tidal bulge on the earth below. Incidentally, it also lifts the land (creating the so-called "earth tides"): the earth's surface lifts approximately 50 cm over the equator as the moon passes overhead. The World Trade Center in New York City rises 36 cm during an overhead lunar transit, and when the moon first rises in the east the two buildings tilt 5 cm toward it, and as the moon sets they lean 5 cm to the west!

Gravitational pull explains the tidal bulge under the moon, but there is another bulge, i.e., another high tide, exactly on the opposite side of the earth. It is formed by centrifugal force, created as follows: The moon does not actually orbit around the earth. The earth and moon *translate* around a point between them. If these two bodies were of equal mass the nodal point would be out in space half way between them. But the earth's mass is 81.5 times greater than that of the moon, and this ratio places their common-mass center inside the earth, about three-fourths of an earth-radius (approximately 4500 km) from our planet's center in the direction of the moon. Thus, a centrifugal force is generated that "throws up" the second high tide on the side of the earth away from the moon. The earth

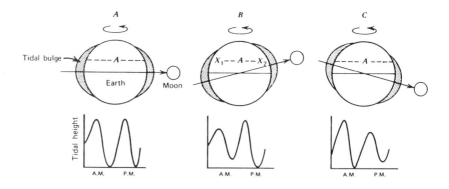

Figure 1-3 The generation of the semidiurnal inequality in tidal amplitude. During the span of one month the moon's declination can range between 28.5° north and south of the equator. When the moon is over the equator (as it is twice each month), both tides during the lunar day at a particular latitude — say at A on the figure — are equal in amplitude (Condition A). When the moon is north of the equator (Condition B), the tidal bulge is greater in the northern hemisphere when the moon is overhead, but on the other side of the earth the deepest part of the high tide is in the southern hemisphere. At this time, at longitude X_1, the shoreline at latitude A is inundated only by the edge of the tidal bulge and experiences a lower high tide than it does 12.4 h later when it is facing the moon (position X_2). The same semidiurnal inequality in tidal depth is created when the moon is south of the equator (Condition C) (Palmer, 1974).

rotates on its own axis under these bulges, thus creating the high to low tidal changes as antipodal longitudes pass under the moon. Because we humans use solar day/night cycles to routinize our lives, we thus recognize the tides as coming and going about 51 min later each day, and each tidal cycle as lasting on average, 12.42 h.

The heights of the tides tend to vary in a regular way. There are several reasons. It takes the moon a month to complete each orbit around the earth, and during this time it does not remain above just a single earth latitude. Instead, it gradually "wanders" first northward, sometimes assuming a final declination of as much as north latitude, and then travels to 28°30′ south latitude, and then returns back north again, completing its migration every 27.32 days (Fig. 1-3). Studying Fig. 1-3A, it should be clear that during the to-and-fro roving, while the moon is directly over the equator (every 13. 66 days), the depths of the two daily high tides at the latitude marked **A** are equal. But when the moon is at its northernmost, or southernmost, declination (Fig. 1-3B and C) the high tides at **A** are not equal, having entered a pattern called the **semidiurnal inequality**.

Now I shall introduce the sun's gravitational attraction into the tidal equation. The mass of the sun is 27×10^6 times greater than the moon, but it is 389 times farther away, so it plays a lesser (45%), but still significant, role in the tidal pattern. When sun, moon, and earth are all in

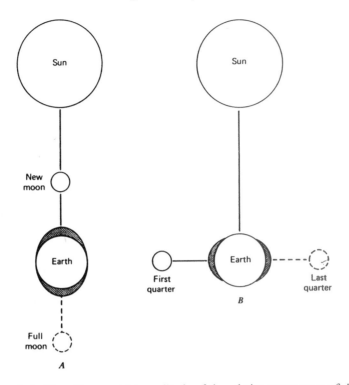

Figure 1-4   The effect on tidal amplitude of the relative movements of the moon, earth, and sun. A: When the three bodies are all in a line, the gravitational forces of the sun and moon combine and cause **spring tides**. B: **Neap tides** are produced when the sun and moon form right-angles with the earth (Palmer, 1974).

a line (the **syzygy**; the times of new and full moon) (Fig. 1-4A), the gravitation of the sun and moon combine to create the highest high tides and lowest low tides of the fortnight. Another way of saying the same thing is that the tidal *range* is greatest at these times. These are called the **spring tides**, and occur twice each month around the times of new and full moon (the spring used here has nothing to do with the season, it comes from the German *springen*, to jump up). When the sun, moon, and earth align themselves as the points of a right-angle (Fig. 1-4B), the sun and moon no longer pull together and the tidal range is smaller. These are called the **neap tides**, and take place at the first and last quarters (the **quadratures**) of the moon's changing phases. The alternations of spring and neap tides thus repeat on a fortnightly basis (Fig. 1-5).

Between the times of the spring and neap tides, another important effect of solar gravitation is to "prime" or "lag" the timing of the tides. Returning again to Fig. 1-4B, the orientation of the sun and moon around the days of first quarter is such that the sun's pull causes the high tides to occur before (*priming*) the moon reaches its zenith, and on the days around

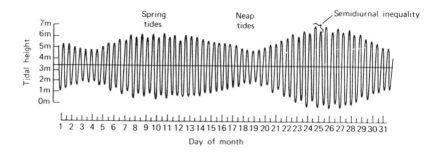

Figure 1-5 An example of the fortnightly alternation of spring and neap tides, and the waxing and waning of the semidiurnal inequality of tidal amplitude (Palmer, 1974).

the last quarter, the sun's gravitation holds back the tides (causing them to *lag*) so that they occur after the moon has reached its zenith.

As a brief aside, let us consider for a moment what effects the semidiurnal inequality and the spring-to-neap amplitude changes have on intertidal dwellers. In each case we will assume animals fixed to the substratum, or living in permanent burrows, or at a preferred level of the shoreline. Living half-way between the upper and lower levels reached by the tides, means that one would be flooded and uncovered by both tides each day. But living at the highest level means that one would be covered only by the spring tides, and/or by the day's major peak when exposed to pronounced semidiurnal unequal tides (Figs. 1-3; 1-4). Living very low in the intertidal means that organisms would be exposed to air only during the spring tides.

Back to the tides. The above describes several major geophysical modifications to the form of the tides. If one wished to make the best possible predictions of future tidal patterns, these primary factors could be combined with 387 other harmonic, partial-tide components to reach that goal. Usually, however, only the four mentioned above are employed in published tables giving tidal schedules. Using one such table, I have plotted a month of tidal data in a special way for Fig. 1-6A; here the data are expressed as tide-to-tide *deviations* from the average tidal period of 12.4 h — the interval routinely used to describe the frequency of the tides. The figure is based on data for the main habitat (near the Marine Biological Laboratory at Woods Hole, Massachusetts, USA) where I collect crabs for much of my work. This kind of presentation allows one to see the incredible tide-to-tide variation in period length that *normally* takes place here. Note that on day 19 the first tide had a period of 11 h 4 min, while the next tide was 13 h and 58 min — a difference of 2 h and 54 min. Note also that groups of days of extreme variability alternate with groups of days that trend closer to the 12.4 h average.

There are more variables. Not only can wide hemiday swings occur

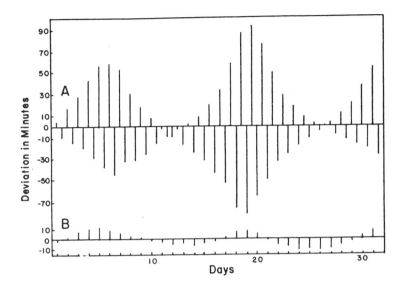

Figure 1-6 A one-month example of the *circa* nature of the period of the tides at a major collecting site near the Marine Biological Laboratory at Woods Hole. A: The deviations in minutes (vertical bars) from a mean period length of 12 h 25 min (represented by the horizontal line at zero) of consecutive tidal-cycle lengths. The most extreme deviations for the month are seen on day 19 where the first tide has a period of 11 h 4 min, while the second one that day has a length of 13 h 58 min — a whopping difference of 2 h 54 min. How could any clock, man-made or living, running at a basic period of 12.4 h, be of any use in a situation like this? B: The deviations in minutes from a period length of 24.8 h (represented again by the horizontal zero line), of one low tide and its repeat the next day. The change in period lengths of these tidal intervals are only 10% of those described in A (Palmer, 1991).

in the fundamental period length of the tides, additional variations are caused by aperiodic meteorological vagaries that affect the timing and amplitude of the tides: barometric pressure and wind are the two most important. The sea acts as an inverted barometer: when the atmospheric pressure drops over a local area the water rises — a fall of 2.5 cm in barometric pressure translates into a rise of about 33.5 cm in sea level. A strong, prolonged onshore wind exerts a push on the water causing it to arrive sooner and "pile up" higher on the shoreline. The opposite is the case with an offshore wind, and even a heavy downpour of rain just offshore can also affect tidal timing.

Some coastlines in the world experience only one tide/lunar day; the northern margin of the Gulf of Mexico, and southeastern Asia are such places. On other coastlines, such as the west coast of the United States, the tides change on a regular basis from one per day to two. Yet another condition contributing to the variability of the tides is the shape of a coastline and the depth of the adjacent water; both factors alter the temporal pattern of the tides. Thus, as will be discussed in greater detail later in the book, the phase of tidal cycles may differ by several hours at spots just a few miles apart along a coastline.

The important take-home message here is that the tidal schedule is nowhere as regular as oversimplifying textbooks teach. While the tide's return is ineluctable, its scheduling appears to have been masterminded by a committee! The 12.4 h period is only an average, and can be arrived at only by combining many days of data. The fact is, that the ocean tidal period is itself really *circa*tidal! And that fact is of prime importance to the subject of organismic, tide-associated rhythms, as will be described at the end of this chapter.

Some authors feel it is *de rigueur* to add a pithy quote or two from a well-known scintillating author. Sometimes their epigrammatic choices are even apropos. I seldom feel that need, but know one that is almost exactly appropriate:

> Time is more complex near the sea than in any other place, for in addition to the circling of the sun and the turning of the seasons, the waves beat out the passage of time on the rocks and the tides rise and fall as a great clepsydra.
>
> John Steinbeck,
> *Tortilla Flat*
> *Amen.* J. Palmer (looking at raw data)

Further, and more detailed, information about tides is found in fine sources such as, Defant (1958), Wylie (1979), and Redfield (1980).

The adaptive significance of a biological clock lies in its ability to alert its owner to the impending periodic changes that will take place in the environment. For instance, a robin's very precise clock awakens its owner and sends it foraging at dawn. An earthworm with an imprecise clock stays out too long and thus becomes the robin's breakfast. Natural selection culls out "late" worm clocks and selects for "early" robin timepieces — thus maintaining the precision of living solar-day timepieces. At the mid-latitudes, the daily differences in the times of sunrise are exceedingly small, changing by only a few minutes each day as the seasons progress. Because organismic solar-day clocks adjust their owners to the precise periodicity of the day/night cycle, solar-day clocks are in general quite accurate. On the other hand, the ebb and flow of the tides are wildly erratic, so much so it would seem that there would be little selection pressure even to evolve any sort of tidal clock, let alone an accurate one (Palmer, 1989). Therefore, unsurprisingly, organismic persistent tide-associated rhythms are the noisiest known, and the small "pendulum" of clockwatchers (how's that for a collective) dedicated to their study, are some of the most haggard in chronobiology. Until one achieves tenure, it would be best not to enter this branch of the field.

Literature Cited

Binkley, S.A. 1990. *Clockwork Sparrow: Time, Clocks, and Calendars in Biological Organisms*. Prentice Hall, New Jersey.

Bünning, E. 1973. *The Physiological Clock*. Third edition. Springer-Verlag, New York.

DeCoursey, P.J. 1983. Biological timing. In:*The Biology of Crustacea. Vol. 7*, Vernberg, F.J. and Vernberg, W.B. (Eds), pp. 107–162. Academic Press, San Diego.

Defant, A. 1958. *Ebb and Flow. The Tides of Earth, Air, and Water*. University of Michigan Press, Ann Arbor.

Edmunds, L.N. 1988. *Cellular and Molecular Bases of Biological Clocks*. Springer-Verlag, New York.

Gwinner, E. 1986. *Circannual Rhythms*. Springer-Verlag, Berlin.

Hastings, J.W., Rusak, B. and Boulos, Z. 1991. Circadian rhythms: the physiology of biological timing. In: Prosser, C.L. (ed.), *Neural and Integrative Animal Physiology*, pp. 435–546. Wiley-Liss, New York.

Moore-Ede, M.c., Sulzman, F.M.and Fuller. C.A. 1982.*The Clocks That Time Us*. Harvard University Prees, Cambridge, MA.

Palmer, J.D. 1973. Tidal rhythms: the clock control of the rhythmic physiology of marine organisms. *Biol. Rev.*, 48: 377–418.

Palmer, J.D. 1974. *Biological Clocks in Marine Organisms: The Control of Physiological and Behavioral Tidal Rhythms*. John Wiley & Sons, New York.

Palmer, J.D. 1976. *An Introduction to Biological Rhythms*. Academic Press, San Diego.

Palmer, J.D. 1989. Comparative studies of tidal rhythms. VII. The circalunidian locomotor rhythm of the brackish-water fiddler crab *Uca minax*. *Mar. Behav. Physiol.*, 14: 129–143.

Palmer, J.D. 1991. Contributions made to chronobiology by studies of fiddler crab rhythms. *Chronobiol. Int.*, 8: 110–130.

Pengelley, E.T. 1975. *Circannual Clocks: Annual Biological Rhythms*. Academic Press, San Diego.

Redfield, A.C. 1980. *Introduction to Tides*. Marine Science International, Massachusetts.

Sweeney, B.M. 1987. *Rhythmic Phenomena in Plants*. Academic Press, San Diego.

Wylie, F.E. 1979. *Tides and the Pull of the Moon*. Stephen Greene Press, Brattleboro, Vermont.

2

Time-Series Analysis

As described in the last chapter the ebb and flow of the tides, while clearly periodic, can follow frenetic frequencies on a day-to-day basis. Thus, there has not been a strong selection pressure for the evolution of precise living clocks in intertidal dwellers. Just how accurate the timepiece of one of these shore dwellers may be is a function of the individual, and/or the species under study. The clocks range from the rare temporal virtuoso, to those in animals that give no overt indication of possessing a timepiece — virtual temporal ciphers. Depending on the species, it is not uncommon to find 50% of a sample in the last category. Cases in which this occurs can result from creating laboratory constant conditions that are unsuitable for the expression of rhythms (meaning that the clocks stop or the coupling severs). Alternatively, the occurrence of these temporal ciphers may be due to the fact that the tidal environment of this group is so erratic that animals can survive adequately without possessing a decent clock.

Thus, most often one has only noisy rhythms with which to work, and this problem must have become immediately apparent to the first investigators. It was solved, somewhat, by lumping together the hour- by-hour responses of several animals. Fig. 2-1, Part I, is a result of such lumping; it shows the combined responses of the spontaneous locomotor activity of 10 fiddler crabs (*Uca pugnax*). The animal's display is far from ideal, but on most days one can see two peaks of activity, and follow these

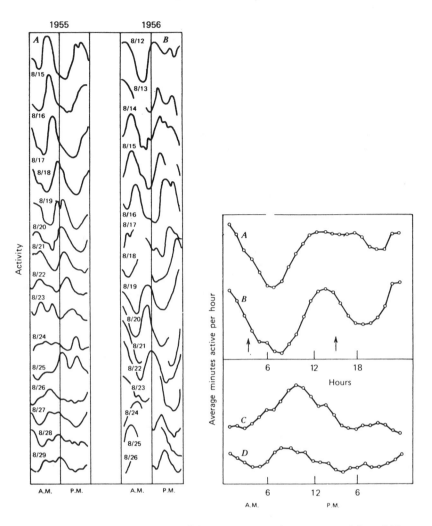

Figure 2-1 Group responses of the spontaneous locomotor activity of 10 fiddler crabs (*Uca pugnax*) maintained in a constant light intensity of <2 ft. can., and a constant temperature of 23°C. Part I (on left) displays replicate observations from two successive years. These data are expressed as 3-h moving averages. Part II, A (on right) illustrates form-estimates for the data in Part I. Unfortunately, the curves are inaccurate because they were constructed using the wrong period estimates. In Part II, B illustrates form estimates of the same data, expressed as a solar-day alignment (Bennett *et al.*, 1957).

maxima's movements to a later time on successive days — just what would be expected of a biological rhythm with a tidal-frequency. Actually, the group data are even more ragged than they appear, because they have been smoothed using a 3 h moving average. The use of this technique to gussy up noisy data has been criticized (Cole, 1957; Enright, 1965) because it is known sometimes to create cycles in data that are truly acyclic. That concern can be real, but not in this case: employing a 3 h slide will not create a 12.4 h period (Dowse & Ringo, 1989). It must be remembered also that this was pioneering work. It came from Frank Brown's laboratory at the Marine Biological Laboratory in Woods Hole a long time ago (Brown *et al.*, 1956; Bennett *et al.*, 1957). Back in those trail-breaking days, workers were unaware of some of the pitfalls of lumping data: they were incognizant that some of the crabs in their sample were, in all likelihood, arrhythmic; and that undoubtedly those that were rhythmic had adopted *different* **circatidal** periods. The latter is probably made manifest by the demise of the curves with time: each study begins with precise, two-peaks/day displays, but then noisiness and a diminution of amplitude begins as individuals with different circa-periods drift out of phase with one another and begin to cancel the waveform in the group's output. That is easy to see with hindsight, but, of course, because it was unexpected back then, it was just accepted as part of a still not fully understood decay of response.

To smooth out the rough edges even more, a type of data-manipulation was adopted that was patterned after a technique borrowed from Chapman and Bartels (1940), two astrophysicists searching for periodicities in geophysical events: All of the data were reduced to a single curve. The activity data were first realigned to a period of 24.8 h, so that all the peaks and valleys lined up when consecutive days were plotted one beneath the other. Then, all the values in the first column were summed and an average calculated, the same was done to the second column, the third, and so on. This reduced all the data in a set to a single, mean-hourly **form-estimate** curve (Fig. 2-1, Part II, A and B). Modifications of this technique are still often used by chronobiologists because they are quite useful in some applications. But in the case at hand (the fiddler crab data) the results are wrong. At the time they were made, the investigators were strong adherents to the exogenous-timing hypothesis of biological rhythms. This hypothesis stated that living clocks received their timing information from periodic geophysical events that were energetic enough to penetrate into otherwise, or "so-called," constant conditions in the laboratory (for an evolutionary overview of this hypothesis see Brown, 1958, 1983; Brown *et al.*, 1970; Webb, 1990). Thus, the activity data were treated as a bimodal rhythm with a period of 24.8 h — the interval of the lunar day. Actually, the group periods had become slightly longer — i.e., *circa*tidal — meaning their alignment to 24.8 h was wrong, as was the resulting curve. Thus, we have lesson one: The exact period of a rhythm must be known before data can be reduced to a representative form-estimate curve.

Fiddler crabs are also exposed to the day/night cycle, and are known to undergo a solar-day rhythm in color change (Brown *et al.*, 1953), so Bennett and her colleagues (1957) next wondered if there might also be a solar-day component in spontaneous locomotion. To extract such an entity — if it existed — they examined 29-day strings of continuous data.

By using this number, they reckoned that because tidal peaks occurred 50 min later each day, each peak would have scanned the 24 h of a day exactly once in 29 days, thus randomizing and eliminating themselves by cancellation. Whatever remnant was left would be a solar-day component. Form estimates of their finding are seen in Fig. 2-1, Part II, C and D, and indicate an activity peak in the forenoon with an amplitude only about a third of that of the tidal component.

Years later, Enright (1965) challenged the existence of the 24-h component, and also the interpretation that the period of the tidal component was exactly 24.8 h. His point was sound: one cannot choose to look only for some expected period, instead one must look for "all possible" periods in a time-series if one hopes to learn what actual periodicity, if any, is present. He chose the technique of periodogram analysis (which will be the next topic of this chapter) and reanalyzed the Bennett *et al.* (1957) data. He found no indication of a solar-day component; he also found that the period was slightly longer than 12.4 hours. Since then another group has re-examined the same data with an even more powerful technique (Maximum Entropy Spectral Analysis, MESA) confirming the circatidal period, and finding no evidence of a 24-h component (Dowse & Ringo, 1989).

Since the completion of this work on fiddler crab rhythms, carried out many years ago, several sophisticated statistical techniques have come into vogue in the attempt to decipher the noisy data of tidal rhythms. Each has its strong and weak points. The chief programs used are: periodogram analysis; autocorrelation; Biomedical & Dental Package, Univariate and Bivariate Spectral Analysis (BMDP-1T); and MESA. In a nutshell, all search for and report periodicities present in a data set, and two of the programs even indicate which cycles found are statistically significant.

As a measure of each technique's usefulness, we will undertake a comparative study. To level the playing field, I have created several different sets of model data, each characteristic of an organismic tide-associated rhythm except that they are ideal: they have precise periods and forms, and are noise free. All the models are constructed of 1-h bins, and consist (except where noted) of 13-day "studies".

Tidal Rhythm Model

The first model is designed to mimic the archetypal tidal rhythm: two peaks/lunar day (in these models the lunar day has been stretched by 70 min to 26 h) of identical form and amplitude separated by 13 h. Obviously, a useful technique will identify the 13-h period.

The periodogram program used is a Williams and Naylor (1978) modification of the Whittaker–Robinson (1944) method. Built into the program is a "periodogram statistic" (Hastings, 1981) that is supposed to give an estimate of the significance of periods identified, but this latter program has been found wanting by Harris and Morgan (1983). The

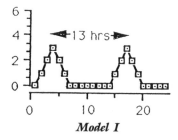

Model I

periodogram found for the above model is seen in Fig. 2-2, and describes a hierarchy of significant periods of: 13 h, 26 h, 39 h, 52 h, and 65 h; the higher values are all *super*multiples of 13.

The BMDP-1T analysis found, in descending order of prominence: 13-h, 4.33-h, and 6.5-h periods; a correct value and two *sub*multiples. This program ranks by prominence the periods that it finds, but does not evaluate their statistical significance.

MESA is a spectral analysis technique that applies an autoregressive model to the data vector, and subsequently extracts information on rhythmicity by Fourier analysis (Ables, 1974; Dowse & Ringo, 1989). It identified major power at 13.1 h, and offers a weak indication of a 6.4-h period (Fig. 2-3). The latter peak is often reported (it is called the "period from Hell" in my lab) because the technique tends to produce low-amplitude near-submultiples when the waveform of the autocorrelation function deviates excessively from a sinusoidal wave (Blackman & Tukey, 1959).

Autocorrelation is especially useful in the study of biological rhythms (Mercer, 1960; Chatfield, 1981), and it clearly indicated the 13-h period (Fig. 2-4).

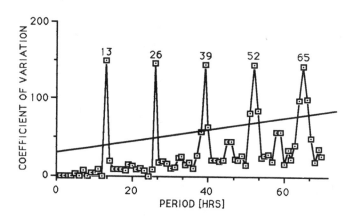

Figure 2-2 A periodogram of the data in Model I. The diagonal line signifies the 95% confidence interval, which is surpassed maximally by the peak at 13 h, and gradually less so for the ascending supermultiples to the right of that spike (Palmer *et al.*, 1994).

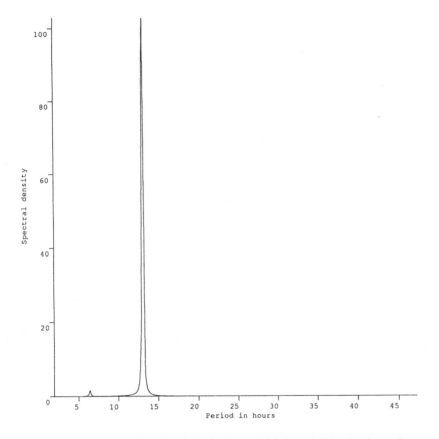

Figure 2-3 A MESA plot clearly indicating a 13.1-h period in the data of Model I. The spurious "blip" at 6.4 h is a common, unwanted, non-significant contaminant (Palmer *et al.*, 1994).

All four programs gave 13 h as their primary answer, but periodogram and BMDP-IT added ambiguity to their output because of the super- and submultiples, respectively, that they also volunteered. The MESA plot, certainly the easiest to read, added the confusion of the secondary peak at 6.4 h. The latter can be identified as artifact by subjecting the same data to autocorrelation (which is always used in conjunction with MESA).

Subtle Dual Period Model

Next, Model I was modified by changing the form of one of the peaks (a common manifestation in persisting rhythms). After the last example, it would be expected that the 13-h period will be found, but, because of the alternating form of the peaks, will a 26-h interval also be identified?

The periodogram reported: 13 h, 26 h, 39 h, 52 h, and 65 h. Here, then, is a problem: Because 26 is a supermultiple of 13 (as seen in the

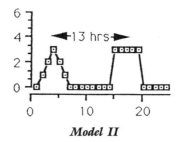

Model II

analysis of Model I), it cannot be concluded that the technique actually identified a 26-h period. BMDP-1T gave: 13 h, 6.5 h, 26 h, <u>3.714 h</u>, 3.25 h, 2.167 h, and <u>2.889 h</u> periods. The two underlined values are submultiple of 26, but not of 13, so thus indicate that the 26-h period in

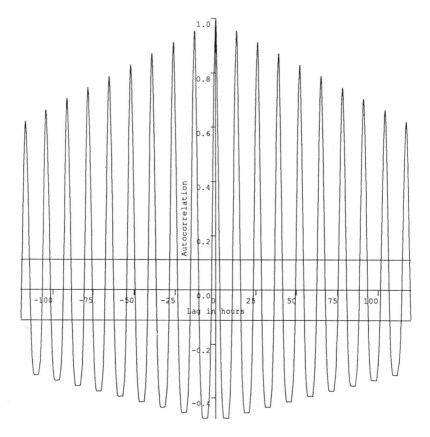

Figure 2-4 A correlogram of Model I data. The spikes to the right and left of the centrally placed ordinate line, fall exactly 13 h apart. The horizontal lines above and below the abscissa represent the 95% confidence intervals. Note that there is no indication here of the 6.4 h period seen in the MESA plot in Fig. 2.3, confirming that spike as just a nuisance (Palmer *et al.*, 1994).

the string of answers has been identified as real (unlike the periodogram output, where it is just a supermultiple of the 13).

MESA shows power at 13.1 h and 25.2 h (as cynics say, "close enough for government work"), and again that annoying pimple at 6.4 h.

Autocorrelation indicated precise periods of 13 h and 25.9 h.

Combined Tidal and Daily Rhythm Model

Intertidal dwellers are also exposed to the day/night cycle and have thus been found to display both basic tidal and basic solar-day periods in a single function such as locomotion (Naylor, 1958; Palmer, 1967, 1974, 1990a). This situation was thus modeled:

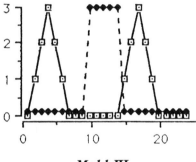

Model III

The same 13-h periodicity used in the first model is copied here, and added to it is a square-wave peak of equal amplitude that is repeated every 24 h. The latter is indicated by a dashed line and diamond-shaped hourly points.

Periodogram analysis produced a family of peaks rising above the 95% confidence line, located at: 24 h, 13 h, **48** h, 26 h, 39 h and **12** h. The underlined values are related as supermultiples of 13, and the bold values as sub- or supermultiples of 24.

BMDP-1T reported 13 h, 22.3 h, 26 h, 6.5 h, 8 h, 6 h, and 4 h. Some explanation is required here. In the higher period ranges, BMDP-IT — with this amount of data collected in 1-hour bins — loses resolving power, so the intervals between the periods it reports increase. In the 24-h range, it indicates only values of 22.3 h and 26 h, and both numbers are given equal weight by the program. This means that the real value lies somewhere between this straddle. However, the 8 h, 6 h, and 4 h submultiples in the output signify the existence of the 24 h period. The underlined 6.5 is a submultiple of 13.

MESA indicated the greatest power at 13.1 h and 24.2 h. The latter value, because the breadth of the 24-h peak, greatly overshadowed the 13.1 one.

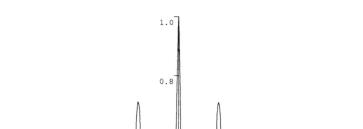

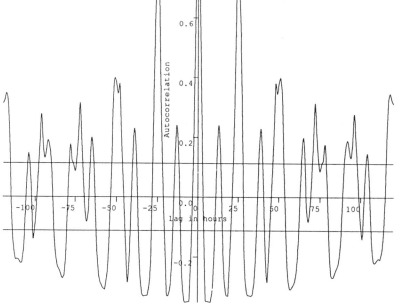

Figure 2-5 A correlogram of the Model III data, containing both "tidal" and "daily" intervals. The first spike to the right of the ordinate represents the 13-h period. The next one to the right of it is a fusion of the 13-h and 24-h periods creating a spike that is relatively taller and wider. The next spike represents the other 13-h peak. And the next is another fusion of the two periods; but here, because 24 is not an exact supermultiple of 13, the peaks have separated enough to make a double-headed display. This division becomes greater from here on (Palmer *et al.*, 1994).

The autocorrelation output is difficult to interpret at first glance. Viewing Fig. 2-5, in the broad middle region of the plot every other 13-h spike must be displayed in virtually the same spot as the 24-h one. This combination does, however, produce single, taller, fatter spikes. But with the fourth spike to the right or left of center, one can see the 24-h peak begin to separate from the 13-h one: it is expressed as a spikelet perched on the top left shoulder of that 13-h spike. Successively thereafter, it emerges as an independent peak. As an aid to interpretation, if you photocopy Fig. 2-4 and superimpose it over Fig. 2-5, the separation becomes obvious.

Tidal Rhythm Combined with an Ultradian Cycle

In the last few years it has been found that short-frequency rhythms — **ultradian cycles** — can accompany tidal rhythms (Dowse & Palmer, 1990; 1992). Finding these ultradian periodicities mixed in with a longer frequency is a challenge for any inferential technique. The following test model was created consisting of a 13-h tidal simulation and a low-amplitude 4-h component:

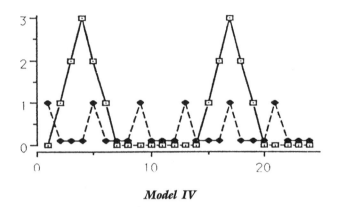

Model IV

Periodogram analysis identified 13 h, 26 h, 39 h, 52 h, 65 h, 4 h, 8 h, 12 h, 16 h, 20 h, etc., periods. While the test, therefore, did identify the 4-h period, so many supermultiples of 13 and 4 were also offered that without any foreknowledge about the contents of the model, interpretation would have been exceedingly difficult. BMDP-1T reported 13 h, 6.5 h, 4 h, 2 h, and 4.33 h, thus also finding the ultradian period. MESA indicated periods at 13.1 h, 4 h, and, as usual, the 6.5-h nuisance. Autocorrelation gave 13 h, and 4-h periods (Palmer *et al.*, 1994).

Peaks That Scan the Day at Different Rates

Palmer and Williams (1986), and several studies after theirs (discussed in Chapter 3), found that in the laboratory the two tidal peaks may not be separated by a constant interval, but instead each has a different, but near lunar-day period. Saying the same thing another way, the two peaks scan the solar day at different rates. This response was simulated by the following model where the peak labelled **e** was given a 26-h period and the **o** peak was given a period of 25-h duration. To increase the resolving power, a 22-day string of data was used.

The periodogram program produced a broadish peak consisting of a fusion of the 25-h and 26-h periods, plus the supermultiples 50 and 52 that indicated indirectly the presence of 25 h and 26 h peaks. BMDP- 1T found 26.4 h, 13.2 h, 8.25 h, 8.8 h, 12 h, 6.6 h and 6.186 h. MESA

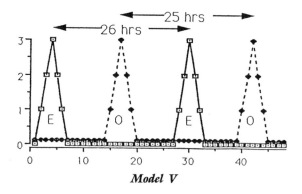

Model V

reported a major peak at 26 h, and some much shorter periods. Autocorrelation indicated a broad peak at 25.5 h, the width of the spike probably representing a fusion of the two periods. Thus, none of the tests produced clear-cut answers.

Conclusions on Models

Periodogram analysis is the most widely used technique in chronobiology, and has been very useful, but does have its limitations. Supermultiples can be very troubling. Suppose you have just collected your first time-series data from an intertidal animal. You wonder if there are basic 12.4-h and 24-h cyclic components in your data, and the periodogram you produced lists: 13.1 h, 26.2 h, etc., periods. Rhythms tend to become *circa* in constant conditions; so are you dealing with a tidal rhythm that has stretched to 13.1 h, and a daily component that increased to 26.2 h, or is the 26.2 just a supermultiple of 13.1? There is no way of knowing using only periodogram analysis.

The Williams and Naylor version of the program only measures period lengths in exact hourly intervals if the data have been collected in one-hour bins, and can thus only approximate the length of important cycles such as the 12.4 h one.

The periodogram statistic can be of help in interpretation, but it can also produce deceptive results. One example was seen with the model containing an ultradian period. Periodogram analysis found the 4-h period, but ranked it sixth in significance.

Probably the program's greatest weakness is its difficulty in handling noisy data — the major bugaboo in the study of many tidal rhythms. This is inevitable since the program's backdone depends on the *variance* of a series of form estimates. Dowse and Ringo (1991) have provided us with a nice comparative example, showing this weakness. They created a 55-day string of data containing two sinusoids of equal amplitude, one with a period of 24 h and the other of 24.8 h. To this they then added 80% random noise, and subjected the conglomeration to periodogram and

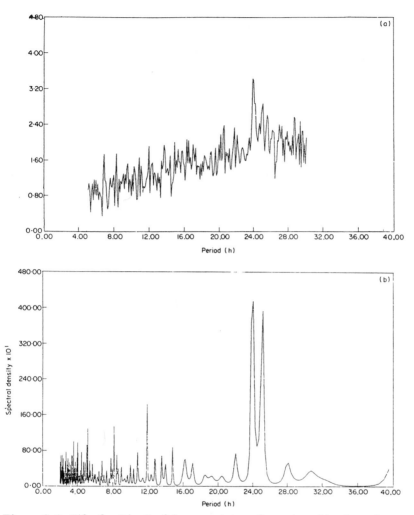

Figure 2-6 Fifty-five "days" of data, consisting of two sinusoids of equal amplitude, one with a 24-h period and the other with a 24.8-h period. These two cycles were then distorted by adding 80% noise. A periodogram (a), and a MESA plot (b) of this artificially generated data are shown. Note that in the periodogram the spectral peaks are almost completely obscured by the noise spectrum, but are conspicuously flaunted in the MESA plot (Dowse & Ringo, 1989).

MESA analysis. Figure 2-6 presents the results. Note that in the former, the spectral peaks are almost completely obscured by the noise spectrum, while MESA virtually "shouts" the presence of the solar- and lunar-daily peaks.

Lastly, the period lengths of rhythms often change spontaneously in constant conditions. The periodogram output is essentially a mean of an entire data set, meaning that a period change goes unnoticed.

BMDP-1T is quite sensitive when used with sparse strings of data containing short periods, such as tidal and ultradian rhythms. To resolve daily rhythms, however, a greater number of days is required. As with periodograms, BMDP-1T produces harmonics, but unlike the former, generates mainly submultiples; and this can be quite an obstacle when searching data for the presence of ultradian cycles. It is quite useful that its output ranks the importance of the frequencies it finds, but it does not indicate the statistical significance of its discoveries.

MESA is a newcomer to chronobiology, and has proven itself to be very useful. It has a tendency to produce low-amplitude, near-submultiples, but their reality is easily checked with autocorrelation. In addition to its accuracy, it is unquestionably the easiest to read of the four programs that are compared here. It comes accessorized with a variety of filters that can limit a search to, or enhance the resolution of, special segments of the periodic spectrum. And, as seen in Fig. 2-6, MESA is an excellent way to examine noisy data. The program does not indicate the significance of periods it finds, but the more power it indicates (i. e., the greater the amplitude of its spikes), the more likely the importance of the spikes. In practice, it is always employed in combination with autocorrelation. (For a copy of the MESA program, e-mail: dowse@maine.maine.edu)

Autocorrelation is a powerful tool for ferreting out periodicities. In addition to finding them, its output displays 95% confidence intervals. Its greatest drawback is the difficulty of interpreting the plots it produces when there are two or more near frequencies and/or a great deal of noise. Using it in partnership with MESA helps remedy this because MESA clearly identifies period lengths that should then be looked for on a correlogram.

The resolving abilities of these four methods on other model simulations, and on organismic data, are found in Palmer *et al.* (1994).

Array Analysis and Compact Plots

Unlike the above programs designed for serious propeller heads, array analysis (Palmer, 1967) is a simple, very useful way to analyze noisy data. The procedure is usually carried out with kludge on a desk computer; but pencil, paper and a grade-school arithmetic ability are all that are required. Yet, it is often the only analysis needed for the study of tidal rhythms. Here is how it is done. Starting with the data collected on the first day, add the 24-hourly values together and calculate the mean (Fig. 2-7). The day is then symbolized as a horizontal line of 24 dashes — one dash for each hour of the day. Then, for each hour in which the activity is equal to or greater than the hourly mean, the dash representing that hour is replaced with a letter **e** for one peak or the letter **o** for the other. An **x** is used to represent an hourly value greater than the mean, but intuitively, that is not felt to be part of a peak and thus considered noise. The treatment condenses a day's worth of data into a simple line of dashes and letters that emphasize

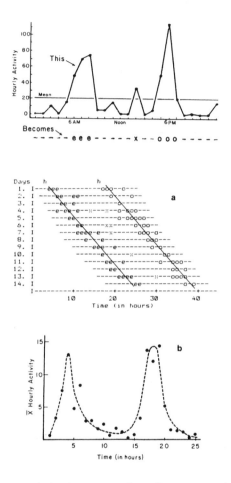

Figure 2-7 Array analysis, the construction of a compact plot, and a form estimate of the result. *Top:* a typical one-day rhythmic display of a crab; a mean has been derived from the hourly activity values and is represented by the horizontal line at 20. That day is then represented as 24 "hourly" dashes. Then the dashes representing those hours in which the activity surpasses the mean hourly line are replaced by letters. *Middle:* subsequent days, thus represented, are plotted one beneath the other and regression lines fitted to the two activity peaks by the method of least squares; the slopes of these lines give the average periods of the peaks — in this case 25.4 h, the rhythm's *circa* period in this set of constant conditions. *Bottom:* one is now in a position to construct a form estimate of the rhythm: in the process all the data are used, not just the values greater than the daily means, to create a mean-hourly representation (Palmer, 1990b).

the major bouts of activity, filters out low-amplitude activity and much of the noise. Successive days are each treated in the same way and plotted one beneath the other. The whole plot is replicated and the copy juxtaposed to the right side of the original; in the center of this plot are seen the two peaks scanning across the solar day. Supernumerary hours can be removed producing a final display like the one seen in the middle of Fig. 2-7. This presentation is befittingly called a "compact plot."

Regression lines, fit by the least squares method, are computed separately for the **e**s and **o**s; the slopes of these lines indicate the average period of each peak, meaning that even though the data were collected on an hourly basis, accurate period estimations to fractions of hours can be made.

Once a period is known, if so desired the data can be aligned to that value and a reliable form-estimate curve produced using all the data, rather than just those values that are allowed to pass through the filter (Fig. 2-7, bottom).

Many advantages are derived from the use of this decoding algorithm. I will list some here, and the usefulness of the procedure will be demonstrated many times in the rest of the book.

1. The technique provides an objective filter that can remove a great deal of noise and minor fluctuations below each daily mean (more or less filtration can be easily created by multiplying the mean by 0.5, 1.5, 2, or whatever, and using only those values larger than the product.

2. The method works just fine even if the period of a rhythm changes spontaneously — as it often does (Fig. 3-3, p. 36).

3. The accuracy of the display is not lessened as a rhythm gradually damps.

4. It easily identifies a split in a peak (Figs 3-4, 3-5, p. 37).

5. It can be superb at distinguishing between two rhythms with different period lengths existing together in a single data set. An example of this is seen in Fig. 2-8, which shows the results of applying array analysis and compact plotting to the dual-period data of Model V, or Fig. 3-25 (p. 59), which is real data. It should also be stressed, that array analysis identified the periods in *all* of the preceding models.

In conclusion, the chronobiologist now has an impressive armada of statistical tools to identify periodicity in time-series data. The four standard mainframe programs are useful in ferreting out low-amplitude, noise-shrouded cyclicity where it may exist. If such discovery is the object of a study, these programs can be especially useful. However, they are less helpful to an *experimental* chronobiologist, because rhythms dependent on their use are often not suitable for meaningful experimentation. As the techniques are not infallible, it is always wise to use two or more methods

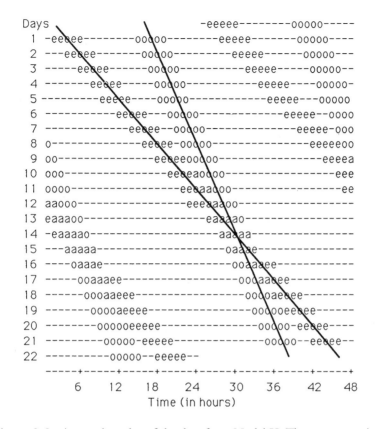

Figure 2-8 A compact plot of the data from Model V. The two regression lines fitted to the **e**s and **o**s estimate the former's period at 26 h, and the latter's at 25 h — exactly correct. The **a**s are used to indicate that both **e**s and **o**s occupy these points (Palmer *et al.*, 1994).

techniques are not infallible, it is always wise to use two or more methods as a means of confirming a result.

Literature Cited

Ables, J. 1974. Maximum entropy spectral analysis. *Astron. Astrophys. Suppl. Series*, 15: 383–393.

Bennett, M.F., Shriner, J. and Brown, R.A. 1957. Persistent tidal cycles of spontaneous motor activity in the fiddler crab, *Uca pugnax. Biol. Bull.*, 122: 267–275.

Blackman, R.B. and Tukey, J.W. 1959. *The Measurement of Power Spectra from the Point of View of Communications Engineering*. Dover, New York.

Brown, F.A. 1958. An exogenous reference clock for persistent, temperature-independent, labile biological rhythms. *Biol. Bull.*, 115: 81–100.

Brown, F.A. 1983. The biological clock phenomenon: exogenous timing hypothesis. *J. Interdiscipl. Cycle Res.*, 14: 137–162.

Brown, F.A., Brown, R.A., Webb, H.M., Bennett, M. and Shriner, J. 1956. A persistent tidal rhythm of locomotor activity in *Uca pugnax. Anat. Rec.*, 125: 613–614.

Brown, F.A., Fingerman, M., Sandeen, M.I. and Webb, H.M. 1953. Persistent diurnal and tidal rhythms in color change in the fiddler crab, *Uca pugnax. J. Exp. Zool.*, 123: 29–60.

Brown, F.A., Hastings, J.W. and Palmer, J.D. 1970. *The biological clock: two views.* Academic Press, San Diego.

Chapman, S. and Bartels, J. 1940. *Geomagnetism.* Clarendon, Oxford.

Chatfield, C. 1981. Short-term rhythms in activity. In: *Aschoff, J. (Ed.), Handbook of Behavioral Neurobiology. V. Biological Rhythms,* pp. 491–498.

Cole, L.C. 1957. Biological clock in the unicorn. *Science,* 125: 874–876.

Dowse, H.B. and Palmer, J.D. 1990. Evidence for ultradian rhythmicity in an intertidal crab. In: Hayes, D.K., Pauly, J. and Reiter, R. (Eds), *Chronobiology: its Role in Clinical Medicine, General Biology, and Agriculture,* pp. 691–697. Wiley-Liss, New York.

Dowse, H.B. and Palmer, J.D. 1992. Comparative studies of intertidal organisms. XI. Ultradian and circalunidian rhythmicity in four species of semiterrestrial, intertidal crabs. *Mar. Behav. Physiol.,* 21: 105–119.

Dowse, H.B. and Ringo, J.M. 1989. The search for hidden periodicities in biological time-series revisited. *J. Theor. Biol.,* 139: 487–515.

Dowse, H.B. and Ringo, J.M. 1991. Comparisons between " periodograms" and spectral analysis: apples are apples after all. *J. Theor. Biol.,* 148: 139–144.

Enright, J.T. 1965. The search for rhythmicity in biological time- series. *J. Theor. Biol.,* 8: 426–468.

Harris, G.J. and Morgan, E. 1983. Estimates of significance in periodogram analysis of damped oscillations in biological time-series. *Beh. Anal. Lts.,* 3: 221–230.

Hastings, M.H. 1981. Semi-lunar variations of endogenous circa-tidal rhythms of activity and respiration in the isopod *Eurydice pulchra. Mar. Ecol. Prog. Ser.,* 4: 85–90.

Mercer, D.M. 1960. Analytical methods for the study of periodic phenomena obscured by random fluctuations. *Cold Spring Harbor Symp. Quant. Biol.,* 25: 75–86.

Naylor, E. 1958. Tidal and diurnal rhythms of locomotory activity in *Carcinus maenas. J. Exp. Biol.,* 35: 602–610.

Palmer, J.D. 1967. Daily and tidal components in the persistent rhythmic activity of the crab, *Sesarma. Nature,* 215: 4–66.

Palmer, J.D. 1974. *Biological Clocks in Marine Organisms: The Control of Physiological and Behavioral Tidal Rhythms.* John Wiley & Sons, New York.

Palmer, J.D. 1990a. Comparative studies of tidal rhythms. X. A dissection of the persistent activity rhythm of the crab, *Sesarma. Mar. Behav. Physiol.,* 17: 177–187.

Palmer, J.D. 1990b. The rhythmic lives of crabs. *BioScience,* 40: 352–358.

Palmer, J.D. and Williams, B.G. 1986. Comparative studies of tidal rhythms. II. The dual clock control of locomotor rhythms of two decapod crustaceans. *Mar. Behav. Physiol.,* 12: 269–278.

Palmer, J.D., Williams, B.G. and Dowse, H. 1994. The statistical analysis of tidal rhythms: tests of the relative effectiveness of five methods using model simulation and actual data. *Mar. Behav. Physiol.*, 24: 165–182.

Webb, H.M. 1990. Biological clocks and the role of subtle geophysical factors. In: Tomassen, G.J., deGraff, W., Knoop, A.A. and Hengeveld, R. (Eds), *Geo-cosmic relations, the earth and its macro-environment*, pp. 56–64. PuDoc, Wageningen, Netherlands.

Whittaker, E. and Robinson, G. 1944. *The Calculus of Observation*. Blackie and Son, Glasgow.

Williams, J.A. and Naylor, E. 1978. A procedure for the assessment of significance of rhythmicity in time-series data. *Int. J. Chronobiol.*, 5: 435–444.

3

A Survey
of Tide-Associated
Rhythms

Fiddler Crabs (genus Uca)

In 1954, a tidal-based oxygen-consumption rhythm was described for two species of fiddler crab, *Uca pugnax* and *U. pugilator* (Brown *et al.*, 1954). On close examination it was seen that the rhythm was probably caused by the animals running-in place on the slippery concave glass of the reaction vessel in the respirator. That possibility was investigated, using the mud fiddler *Uca pugnax*, and the first tidal rhythm in crab locomotion was described (Bennett *et al.*, 1957). As described in Chapter 1, and shown in Fig. 1-1 (p. 6), the revelation was based on the pooled response of ten to 20 crabs, the grouping being used as a crutch to compensate for the usually high noise-to-signal ratio of their display. A convincing rhythm was unveiled, but important information could also remain concealed in the data amalgamation. What was needed next was a careful examination of the temporal responses of *individuals* to a laboratory environment in which all the obvious time cues had been purged. Doing this is not an enviable task because the experimenter finds that many of the animals are arrhythmic; some have rhythms that are accompanied by so much noise that only the use of some of the sophisticated techniques described in the last chapter will separate rhythm from rubbish with any certainty; and sadly, only a

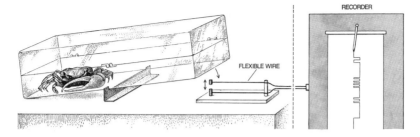

Figure 3-1 A mechanical tipping-pan actograph. A crab, along with enough water to keep the internal environment saturated, is sealed in the pan. As the animal moves from one end to the other the pan teeters back and forth closing and opening the microswitch. Each short migratory trip is thus recorded as a spike on an event recorder outside of the isolation chambers where the actographs reside. Another type of actograph consists of a stationary container with a beam of infrared light transfixing it. Each time the incarcerated animal interrupts the beam, the interference is counted by a dedicated computer out in the laboratory (Palmer, 1975).

small proportion of the subjects have cycles sufficiently clear-cut to be used in interpretation or experimentation. Another special frustration is that after studying literally thousands of individuals, an insight is developed akin to "thinking like a crab," and a lowly human observer becomes certain that he understands what the animal is doing (even though the results are several digits to the left of significance), but knows he could never convince an outsider of the fact. Additionally, becoming decapomorphic does little for one's social life. And then there is the tedium and the extra work of experimenting with large samples of crabs just to find a sufficient number with timepieces precise enough to convince yourself, and the small splinter group of other chronobiologists that work with marine organisms, that your findings are real. In short, one's personality must contain a tincture of fanaticism.

Is tracking *individual* rhythmicity worth it? I think so — after all this is science. Dr John Aldrich at Trinity College, Dublin, makes it clear: "No serious minded investigator would extrapolate the average responses of a group of human prisoners to predict the individual responses of those outside, yet this is the implication of many former laboratory studies" (Aldrich, 1979).

Persistence in this case has paid off, as you will see with the continuation of the fiddler crab chronicle. I will start with the best. Animals are placed individually in actographs, usually home-made versions such as the one shown in Fig. 3-1, and maintained in constant conditions. Then, as the animals trapeze to and fro in their prison, their activity is automatically recorded, and every so often one stumbles onto an animal with a very precise chronometer. Keeping in harmony with today's TV news-readers' hyperbole, my laboratory always identifies such finds as

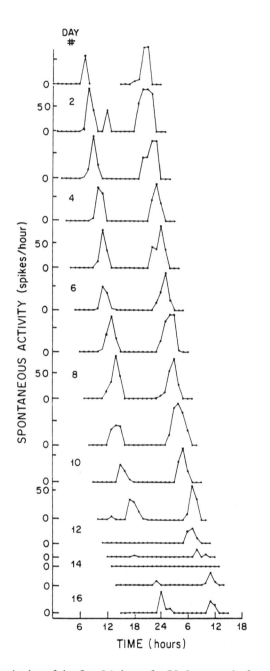

Figure 3-2 A plot of the first 16 days of a 58-day record of spontaneous activity of a mud fiddler crab, *Uca pugnax*. The animal was exposed to constant dim illumination and a temperature of 22°C. Note that the two daily peaks are expressed about 50 min later each day, just as are the tidal movements on the animal's home mudflat. Six hours of the record are missing on the first day due to a mechanical failure of the chart recorder (modified from Palmer, 1989b).

"supercrab." The record of the first 16 days from one of these bravura performances is seen in Fig. 3-2. The rest/activity pattern is approximately what would be seen on the shoreline: running about during low tides and sitting quietly in its burrow during high tides. Note that during the first 11 days the horologue made only one mistake: it inserted a superfluous peak on day 2. Then on days 12–14, one of the peaks disappeared; but this peak returned on day 15 and, by the next day, it had grown back to a respectable size. This turns out to be a common finding, and, because the peak reappears in a phase that would be expected if it had persisted throughout that interval, it suggests that the coupler between clock and locomotion had been temporarily disabled. The peak then persisted for the next 42 days, after which the crab was released (still rhythmic) back in its home territory — hopefully to spread its elite genes widely (Palmer, 1989b).

That the period of the above animal's rhythm tended to match that of the tides is not indicative of what is found in all fiddler crabs. In fact, the periods of most crabs in the laboratory are usually somewhat longer than the average interval of the tides (Barnwell, 1966; Palmer, 1963, 1973): to give a specific example, for 33 *U. pugnax*, kept at 15°C. and in constant darkness, the average period length was 25.05 ± 0.63 h (Palmer, 1988). Nor is the period necessarily constant; it is not unusual for it to change spontaneously one or more times during a stint in constant conditions. The most vertiginous example of this ever recorded is seen in Fig. 3-3. For the first 28 days in constant darkness the crab displayed an extremely long average period of 28.5 h! The period then abruptly shortened to 21.8 h for approximately 5 days, and then lengthened to 26.1 h. This anfractuous exhibit was not the result of some inconstancy in the incubator environment; the other 19 crabs, sitting side-by-side with this one, underwent no simultaneous change in period (Palmer, 1989a). Such spontaneous changes are known also for circadian rhythms (Palmer, 1964b; Pittendrigh & Daan, 1976a).

Let us digress for a paragraph to discuss some terminology. When a rhythm is not entrained it is commonly said to be describing its "natural period." The claim has an interesting history. At one time, chronobiologists bickered over whether a biological clock developed its own period autonomously, or whether even in the laboratory it received its timing information directly from the environment. The two ideas were called the endogenous- and exogenous-clock hypotheses, respectively, and became quite a quodlibet. Those preferring the latter suggested that certain geophysical forces that described 24 h cycles in intensity — such as background radiation, geomagnetism, etc. — which easily penetrated otherwise constant laboratory conditions, could be used by the living clock to signal the passage of 24 h to incarcerated test organisms. The en-dogenous camp denied such a possibility, reasoning that if that was true, the period length of persistent rhythms would be exactly 24 h long, rather than the usual circadian periods exhibited. They felt that the only reason

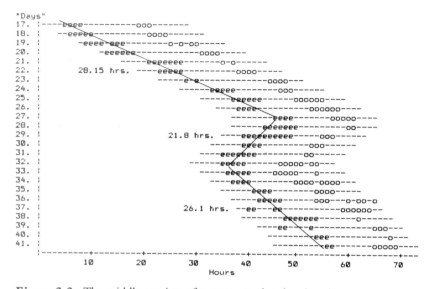

Figure 3-3 The middle portion of a compact plot showing the spontaneous change in frequency of an individual brackish-water fiddler crab, *U. minax.* (See text for details.) The final period was maintained for 26 days until the crab, still rhythmic, was released. The animal was maintained in constant darkness and at 15°C. The cause of the changes is unknown, but less tortuous ones are common (Palmer, 1989a).

24-h rhythms were such was because they were entrained to the day/night cycle. To further inculcate this point into a chronobiological mind-set, they described the non-24-h period produced in the laboratory as the "natural period," the putative rate at which the clock really ran. While their reasoning was sound (and eventually ended the ideological gun-fight between the warring camps), we should question the now common usage of natural period. Re-examine Fig. 3-3; which period would you choose as the natural one? Spontaneous change in period length is very common. Maybe we should be discussing natural period*s*. Furthermore, an experimenter can usually "create" a desired period length in a persistent solar-day rhythm by adjusting the *intensity* of the ambient constant illumination. Thus, to me, it seems rather obvious that what we reflexively call a natural period is really an *un*natural one, in that it is created by the *un*naturalness of constant conditions, rather than being a "pure" expression of the clockworks. Because a kind of Heisenberg uncertainty principle exists here, where the test conditions needed to elucidate a basic period clearly preclude a chance of an accurate answer, it would probably be better to let *natural period* descend from lingo into limbo. End of exegesis; back to the seashore.

Occasionally in the laboratory one of the peaks will spontaneously split in two. Examples of this are seen in Figs 3-4 and 3-5. In Fig. 3-4, after

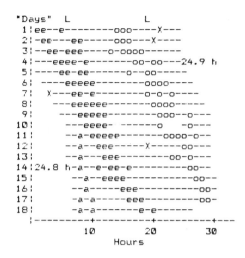

```
"Days"  L                    L
  1 !ee--e--------ooo----X---
  2 !-ee---ee-----ooo---X----
  3 !--ee-eee----o-oooo------
  4 !---eeee-e-------oo-oo---24.9 h
  5 !----ee-ee------o--oo-----
  6 ! ----eeeee-------oooo----
  7 !  X---ee-e--------o-o-o----
  8 !   ---eeeeee-------oooo-----
  9 !   ---eeeee--------ooo--o---
 10 !   ---eeee- ------o   -o---
 11 !   --a-eeeee-------oooo-o--
 12 !   --a-eee-----X-----oo---
 13 !   --a---eee--------oo-o---
 14 !24.8 h-a--e-ee-e--------oo--
 15 !     --a--eeee-----------oo--
 16 !     --a-----eee----------oo-
 17 !    -a-a----eee---------oo-
 18 !    -a-a-------e-e-----
    !---------+---------+---------+---
          10        20        30
                  Hours
```

Figure 3-4 A compact plot of a single *U. pugnax* exposed to constant darkness and 22°C. On day 11, a small branch (indicated by the **a**s) splintered off of the **e** peak and adopted a period of approximately 20 h (Palmer, 1989b).

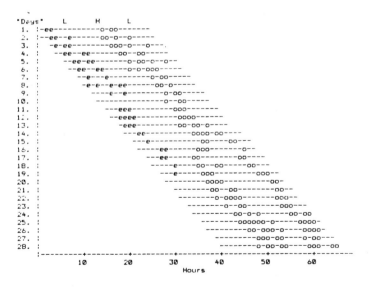

```
"Days"  L        H        L
  1. !-ee----------o-oo------
  2. !--ee--e--------oo-o--o-----
  3. ! -e-ee---------ooo-o---o---.
  4. !  --ee--ee------oo--oo-----
  5. !   --ee-ee------o-oo-o--o--
  6. !   -ee--ee----o-o-ooo-----
  7. !    --e---e---------o-oo-----
  8. !   -e-e--e-ee------oo-o-----
  9. !    ----e--e--------o-oo-----
 10. !     ---------------o--oo-----
 11. !     ---eee---------ooo-------
 12. !     --eeee---------oooo------
 13. !    -eee---------oo-oo-o----
 14. !     ---ee---------oooo-oo----
 15. !      ---e---------oo----oo---
 16. !      -----ee-----ooo-------o--
 17. !      ---ee-----oo--------oo----
 18. !      -----e----oo--oo----oo---
 19. !     ---e-----ooo---------ooo--
 20. !        ---------oooo----------oo-
 21. !       --------oo--oo--------oo--
 22. !       --------o-oooo-------ooo--
 23. !       --------o--oo------ooo---
 24. !       --------oo-o-o------oo-oo
 25. !       -------oooooo-o----oooo-
 26. !       --------oo-ooo-o----oooo-
 27. !       --------ooo-oo----o-oo---
 28. !       -------o-oo-oo---ooo--oo
    !---------+----+---------+------+---------+----+-------
          10        20        30        40        50        60
                         Hours
```

Figure 3-5 The temporal display of a single *U. pugnax* over 28 days while maintained in constant darkness at 15°C. After 13 or 14 days the **o** peak divided, and the two descendants ran parallel to one another 12-h antiphase, at the same period as their parent peak — 25.6 h. Also, after day 19, the **e** peak disappeared, in this case, never to return (Palmer, 1988). (See Fig. 3-6 for further details of the data.)

37

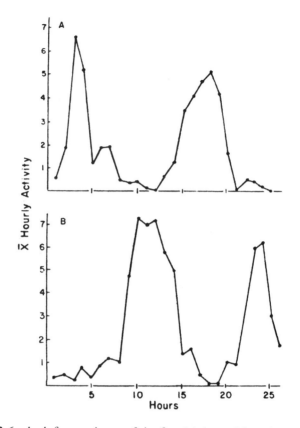

Figure 3-6   A: A form estimate of the first 14 days of data shown in Fig. 3-5, showing the shape and relationship of the **e** and **o** peaks. B: A form estimate for days 19–28, after the **e** peak had disappeared, and the **o** peak had split in two. All the data (not just those values that exceeded the daily, hourly mean) were used in the construction of these two curves. This means of presentation clarifies that there is no trace left of the **e** peak, and that there was a complete division of the **o** peak (Palmer, 1988).

about 10 days in constant conditions, a small fraction of the **e** peak is seen to be cleaved off, assuming a period of 24 h (Palmer, 1989b). The data presented in Fig. 3-5, indicate that in this case it was the **o** peak that fragmented with the two resulting branches running parallel to one another through day 28. Note also that after day 19, the **e** peak was no longer displayed. To further elucidate the results from this crab, two form estimates were made. The upper curve, seen in Fig. 3-6A, is based on the first 14 days of data realigned to 25.6 h. The narrow **e** peak is to the left and the broader **o** peak to the right in the figure. The form estimate in B of that figure, based on days 19–28, shows the total extinction of the **e** peak and the two parallel branches formed by the **o** peak (Palmer, 1988).

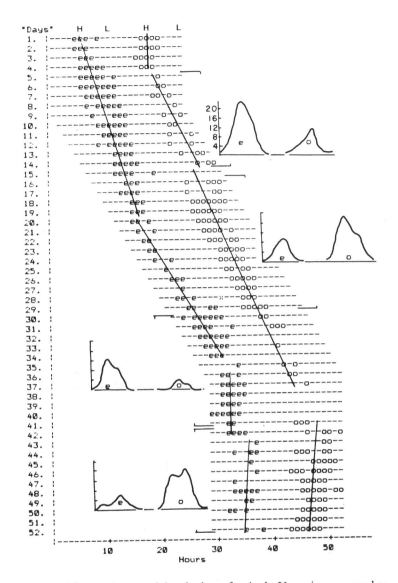

Figure 3-7 The persistent activity rhythm of a single *Uca minax* exposed to 15°C and constant darkness for 52 days. During this long interval there was an alternation of dominance between the **e** and **o** peaks. To illustrate this, form estimates were constructed for days 5–14, 15–29, 30–41, and 42–52, and juxtaposed next to the segment of the data that they represent. The ordinate intervals for all four inserts is the same, with the actual values indicated only on the upper one. Note also the spontaneous changes in period lengths, ending with a rare period of less than 24 h; and the temporary disappearance of the **o** peak after day 37. This crab remained active for more than a fortnight after day 52, but had become arrhythmic when it was finally released back onto its mudflat (Palmer, 1989a).

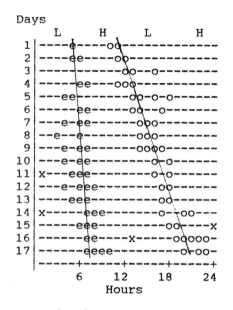

```
Days
         L        H        L        H
   1 | ----e-----oo------------------
   2 | ----ee----oo------------------
   3 | -----|-----oo---o-------------
   4 | ----ee---ooo------------------
   5 | ---ee|-------oo--o-o----------
   6 | -----ee-------oo---o----------
   7 | ---e-ee-------opo--------------
   8 | --e--e--------ooo-------------
   9 | ---e-ee-------oooo------------
  10 | ---e-ee--------o-o-----------
  11 | x---eee--------o-o----------
  12 | ---e-ee-------oo-----------
  13 | ----ee|-------oo-----------
  14 | x-----eee--------o--oo----
  15 | -----eee---------oo----x
  16 | ------ee----x-----oooooo-
  17 | ------eeee---------o-oo--
      -----+-----+-----+-----+
         6       12      18      24
                Hours
```

Figure 3-8 A compact plot of a single *Uca pugilator*, held in constant darkness and at 15°C, showing that the peaks of a tidal rhythm sometimes scan the solar day at different rates. The **e** peak had a period of 24.2 h and the **o** peak, 24.6 h. The two differed significantly: $P < 0.001$ (Palmer, 1988).

Such splitting in *Uca* is seen in about 15–25% of the animals collected, depending on the laboratory conditions used. Peak fragmentation, as will be seen later in this chapter, has been found in other genera as well. It has also been seen in the circadian rhythms of lizards (Underwood, 1977), birds (Gwinner, 1974), and is common in small mammals (Hoffman, 1971; Pittendrigh & Daan, 1976b; Turek et al., 1982).

The rhythm of the brackish-water fiddler, *U. minax*, is depicted in Fig. 3-7; in addition to describing a temporary interval of peak loss, a sudden phase jump on day 43, and some spontaneous alternations in period length, this animal also underwent several modifications of form and amplitude of the **e** and **o** peaks. The changes can be seen in the compact plots, but these are much clarified by the adjacent form-estimate inserts. The details of the construction of the inserts are given in the figure legend. The natural tides in this crab's habitat also alternate in a somewhat similar manner, suggesting a possibility of the extension of the causal relationship between moon and tide, to moon and crab in otherwise constant conditions. As intriguing as this thought might be, the alternations in this animal, and in others describing similar alternations, are not at all in phase or share the same periods with the semidiurnal inequality of the natural tides (Palmer, 1989a).

All of the above findings represent unexpected properties, and are therefore very interesting; but certainly the following discovery was the most surprising and intriguing one of the lot. I say "surprising," because up until this time there was no reason to consider organismic tidal rhythms as other than basically 12.4-h cycles, and "intriguing" because it mandated that we give second thought to our ideas about the tidal clock and couplers. We found that a few animals, when subjected to the unnaturalness of constant conditions, brandished peaks that scanned the solar day at different rates! Fig. 3-8 provides an example supplied by the sand fiddler, *U. pugilator*. The **e** peak had a period of 24.2 h and the **o** peak one of 24.6 h; the difference between the two was highly significant: $P = <0.001$ (Palmer, 1988). It seems unlikely that such a display would occur in the natural setting for two reasons: (1) there the rhythms are entrained by the tides, and (2) should the entrainment break down the nonconformist would be quickly selected out by aquatic predators. But, in the unnatural constancy of the laboratory there are no entrainment dictates, and not even poorly paid graduate students eat *Uca de Jour*, thus mistakes in timing have no repercussions. Checking back through the older literature, I found other examples of this: in Webb's 1971 paper I found it in Fig. 1e (and calculated that the difference was statistically significant), and also in Fig. 8 of Barnwell's 1966 paper. As you will find in the following pages, it has now been discovered also in several other genera.

I will mention briefly here, and in much greater detail in the last chapter, the epiphany resulting from the findings described above. First, the basic conundrum. There has been a long-standing question about these tide-associated rhythms: Should they be considered *uni*modal 12.4-h cycles, that is, is their basic period that of the tides; or would it be more appropriate to consider them *bi*modal 24.8-h rhythms (Palmer, 1974)? The question is stated graphically in the top of Fig. 3-9. How can the fiddler crab data help with a solution? It should be clear that the **e** and **o** peaks can act independently of one another; one might even be tempted to claim that they have "minds of their own." "Clocks of their own" is the claim of the hypothesis educed to explain them. Recognizing that one of the two tidal peaks can disappear while the other does not; that one can split, while the other does not; that the amplitudes and form can differ between the two; and, most importantly, that the two can scan the solar day at different rates, pretty much mandates that the answers to the two questions asked above are "wrong" and "wrong." Instead, it seems likely that each peak is controlled by its own clock, running at a basic period of 24.8 h. The two *uni*modal, *lunar-day* rhythms produced are phase locked together, 180° apart. This idea is illustrated in the lower half of Fig. 3-9, where the phase lock, i.e., the coupling, is represented by a padlock. Most of the time in the lab the coupling remains intact so that the overt expression of the rhythm is indistinguishable from a basic tidal rhythm. But the unnaturalness of constant conditions permits, in some cases, the coupler between clocks to break, allowing the different period lengths to emerge. Or the coupling between one of the clocks and its governed peak breaks so that locomotion is not periodically stimulated and the peak vanishes — temporarily or permanently. Or one clock will, for some

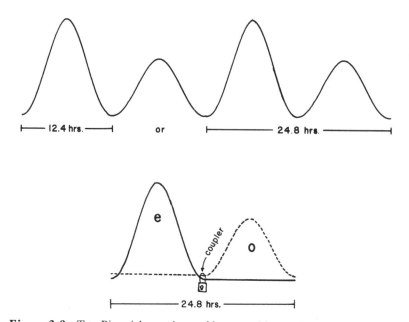

Figure 3-9 *Top:* Pictorial questions asking: are tide-associated rhythms basic unimodal, 12.4-h cycles, or are they fundamental bimodal, 24.8-h cycles? *Bottom:* After deciding "neither," and developing a "circalunidian clock hypothesis" to explain the laboratory results, the illustrated solution is given. There are two unimodal, 24.8-h clock-controlled rhythms that are usually coupled together 180° out of phase (Palmer, 1990b).

unknown reason, produce a bimodal output, creating the overt appearance of a split. Thus each tide-associated peak is controlled by its own lunidian clock that shorten or lengthen somewhat in constant conditions, giving rise to the ultimate copulative, "circalunidian" (Palmer & Williams, 1986b; Palmer, 1990b). To describe the peaks' sovereignty, my colleague Barbara Williams and I refer to the idea as the **circalunidian clock hypothesis** (note the spelling, the *bon mot* mavens in my university's English Department dictate that the underlined letter should be an "i", rather than the "a" that I used back in 1973, so that my neologism conforms to the pre-existing *lunisolar* and *lunitidal* adjectives), to replace the somewhat shopworn "circatidal."

Lastly, fiddler crabs also exhibit ultradian rhythms. By definition these rhythms have periods ranging between approximately 1 and 10 hours. The term may be pronounced "ultra dian," but is usually articulated as "ul **tray** dee an." (If marine chronobiologists were to conform with our landlubber counterparts, we would be studying "sir **cat** ih dal" rhythms: ridiculous.) Many years ago, a hypothesis was developed claiming that *circadian* rhythms were an expression of the combined output of many short-period cellular oscillators coupled together (Goodwin, 1963; Pavlidis, 1971). Ultradian rhythms are the hypothetical output of the more recalcitrant of

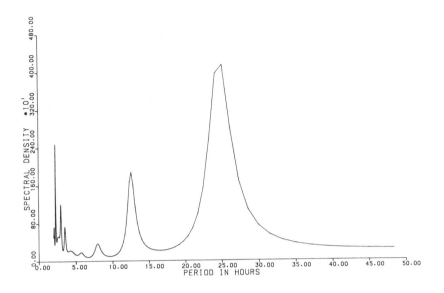

Figure 3-10 A MESA plot of the activity rhythm of a single *Uca pugnax* subjected to 22°C and constant dim illumination. The three major spikes, all deemed significant by autocorrelation, arise at an infradian 2.5 h, and at circatidal and circalunidian intervals (Palmer, 1989b).

these individual oscillators that have broken free of the composite and thus express themselves alone; they have been found associated with circadian cycles (Dowse & Ringo, 1987). Now they have been found to accompany the circalunidian rhythms of *Uca pugnax* also (Palmer, 1989b; Dowse & Palmer, 1990; 1992). To give just one of many examples, the MESA plot seen in Fig. 3–10 indicates not only significant spikes at *circa* tide-associated frequencies, but also an ultradian cycle at 2.5 h.

The Green Shore Crab *(Carcinus maenas)*

Shortly after the discovery of the tidal rhythms in the fiddler crab, Ernest Naylor (1958), then at University College of Swansea, UK, described one in activity for the green (some of them are red!) shore crab. Smaller individuals live in the intertidal zone under rocks during *low* tides, and emerge to forage underwater during high tides — especially during night-time high water. Larger ones hide under rocks at the lowest zone of the intertidal, or remain underwater in the shallow subtidal during low tide, and then migrate inshore with the flood tide and retreat back to their starting point with the ebb (Naylor, 1958; Warman *et al.*, 1993).

Specimens were collected and returned to the laboratory and placed in saturated air or underwater in constant conditions. As seen in Figs 3-11 and 3-12, the group response of several animals described a tidal rhythm that also contained a strong solar-day component, in that the night-time

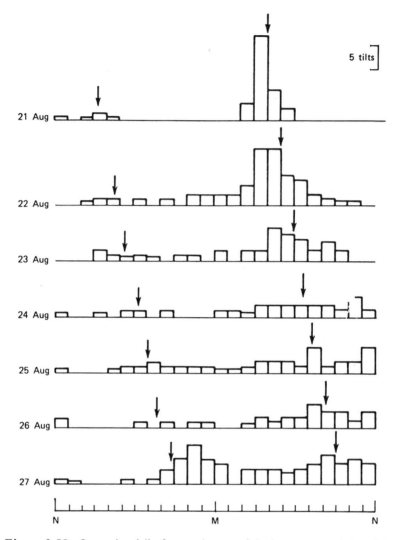

Figure 3-11 Successive daily form estimates of the locomotor activity of three green shore crabs, *Carcinus maenas*, for 7 days in constant dim illumination. The time of day is plotted on the abscissa; the falling arrows signify the times of high tide (Naylor, 1958).

tide-associated peak was significantly heightened. Crabs collected at times when the two low tides occur during the hours of daylight, then show a third peak at night — the unencumbered expression of a nocturnally phased solar-day rhythm. That this is the proper interpretation of the peak is proved by showing that it alone can be entrained to a 24-h light/dark cycle (Naylor, 1958, 1960, 1962; Atkinson & Parsons, 1973). Subsequently, the same laboratory described a tidal rhythm in oxygen consumption (Arudpragasam & Naylor, 1964), and much later a rhythm

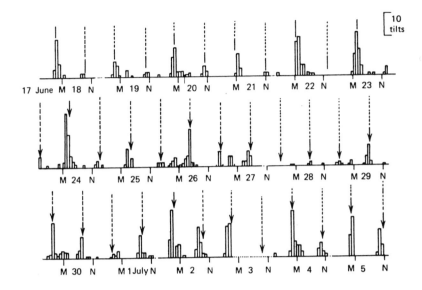

Figure 3-12 The combined locomotor-activity output of 18 *Carcinus maenas* in constant dim light. Each day at 2 p.m., a completely new batch of freshly collected crabs replaced the used ones in the actographs. This experimental design was chosen to demonstrate the night-time enhancement of that tidal peak. The falling arrows again represent the midpoints of high tides (Naylor, 1958).

in blood-sugar concentrations, in which sugar concentrations are 40–45% higher at expected low tides (Williams, 1985). From this humble beginning, and mainly using this crab, Professor Naylor and his students have been very productive, and have discovered most of the fundamentals now known about tidal rhythms. This service has been recognized and Professor Naylor now, quite deservedly, holds the Lloyd Roberts Chair at the University College of North Wales's School of Ocean Sciences.

Naylor (1960) then learned that the tide-associated rhythm was present only if the crabs were exposed directly to the tides before being brought into the laboratory. This elucidation resulted from using crabs dwelling on the lower side of a floating "dock," a location which, of course, precluded the animals from experiencing inundation cycles. In constant conditions these animals manifested only circadian rhythms.

Naylor next studied *Carcinus mediterraneus* from Italy. On the shore boarding the Stazione Zoologica Napoli, the tidal heights are small, only approximately 1.4 feet. These animals also displayed only circadian rhythms in the laboratory (Naylor, 1961). Warman *et al.* (1991) has since confirmed this finding.

After learning of results like these, one might wonder if the tidal frequency has to be learned from the environment before it can be expressed in the laboratory. Barbara Williams, a Naylor postdoctoral

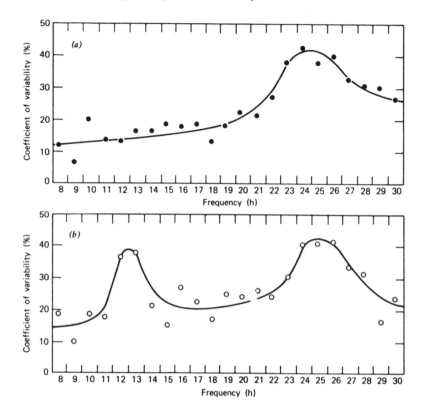

Figure 3-13 Periodogram analysis of the activity rhythm of "home grown" *Carcinus maenas* in constant dim illumination and 15°C. The crabs were raised from the egg in the laboratory in natural day/night cycles and constant temperature, and then placed in actographs in constant conditions. As shown in (a) the rhythm displayed was circadian with a period of 25 h. They were then given a 15-h cold pulse of 4°C and returned to the actographs. As can be seen in (b), that simple treatment initiated a tidal component in their activity (Williams & Naylor, 1967).

student, and my colleague for many years, provided the answer to this question. First she solved the difficult problem of raising green crabs from egg to adult (she clearly has a "green" thumb) in the laboratory. In doing this she used a constant temperature of 15°C, and natural day/night alternations. When her charges reached a size suitable for actograph measurements they were placed in constant conditions where they displayed a distinct circadian (only) activity pattern. Periodogram examination of the output gave no indication of a tidal component in the data (Fig. 3-13a). As will be described later in the book, a single, short-term exposure to cold will reinstitute a tidal frequency lost in animals that have become arrhythmic due to storage in the laboratory (or in the "floating-dock" crabs described above), so she subjected her naïve subjects to one 15-h pulse of

4°C; and then returned them to actographs for more automatic observations. A tidal component had been clearly instilled (Fig. 3-13b), with the first peak set by the time of return to 15°C, and subsequent maxima then repeating at roughly 12.4-h intervals (Naylor, 1963; Williams & Naylor, 1967). Because a tidal interval could not have been learned from one 15-h cold shock, clearly the tidal frequency is innate. The clock was there all the time; it just needed to be "wound up" before it could start running!

Back in 1655, one of the great problems with the newly invented pendulum-based clock was thermal extension of the pendulum arm. On warm days the clock ran slower and on cold days it ran faster as the length of the arm changed. Temperature compensation was finally instilled by using a mercury-filled tube as the weight at the end of the pendulum arm. As rising temperature caused the arm to lengthen, the mercury simultaneously expanded and rose in its tube, thus maintaining the point of swing at a constant radius. Without such *compensation*, a clock stops functioning as a horologue and becomes a difficult-to-read thermometer. The same holds for a living clock, but the problem would be much greater because the clock is based on chemical reactions, which are notorious for their temperature sensitivity, often at least doubling for every 10°C rise in temperature. To be of use to an organism a clock must be temperature compensated. In testing this in *C. maenas*, Naylor (1963), in separate experiments, placed rhythmic subjects that had been maintained at 15°C, into dim constant light and constant temperatures of 10°, 15°, 20°, or 25°C. Over the next six cycles there was no lengthening or shortening of their circatidal periods. Just how this indifference was maintained remains a mystery.

As stated before, although studying only the group response of subjects buries much of the noise, important information can be lost. Barbara Williams therefore took on the Promethean challenge and became an interloper into the individual temporal lives of *Carcinus*. Some of her results follow. After 6 days in constant conditions the mean response of a group of *Carcinus* indicated arrhythmicity, and periodogram analysis confirmed this. Nevertheless, data continued to be gathered and after 3 weeks the activity patterns of *individuals* were scrutinized. The expressions of two of these animals are shown in Fig. 3-14. One crab shows a very clear *c.*27-h rhythm, while the other has what might best be called a trimodal *c.*26-h rhythm. As you can see, a great deal of useful information can remain concealed in data analyzed only as far as a group central tendency (Williams, 1995).

In another study consisting of 54 crabs, Williams (1991), unsurprisingly, found a good deal of variability in the expression of their temporal displays. She collected her animals at specifically selected tide/day-night combinations: (1) when the high tides came in the early afternoon and middle of the night; (2) with the high tides at early evening and just before dawn; and (3) with both high tides in daylight. Assigning a meaning to the results is about as difficult as interpreting the group grope of a bucket

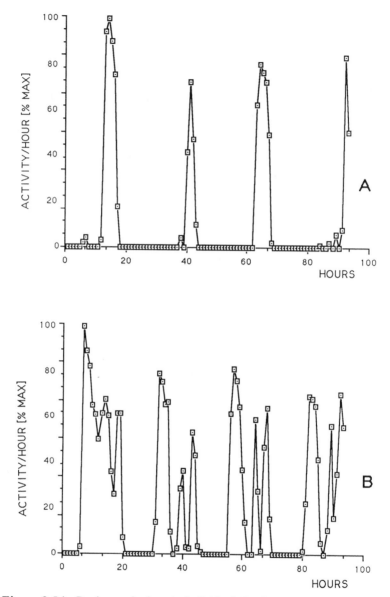

Figure 3-14 Persistent rhythms in individual *Carcinus maenas*. A: a very clear
*c.*27-hour circadian rhythm. B: probably a trimodal *c.*26-hour circadian rhythm
(Williams, 1995).

of crabs scrambling over one another. In the first alignment (1), all 23
animals were active at the time of the expected night-time tide. For half
of them, this was the only activity expressed in constant conditions (Fig.
3-15). The other half underwent a second bout of activity at the time of
expected day-time high tide, but the effort, and thus peak amplitude, was

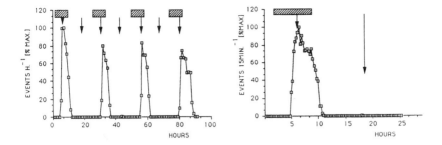

Figure 3-15 Left: The activity display of a single *C. maenas* in constant conditions for 4 days. The falling arrows indicate the expected times of high tides, and the shaded rectangular overhead "clouds" indicate the expected hours of darkness. Note that peaks formed only at *night-time* high tides. Right: A 25.25-h form estimate of the same data (Williams, 1991).

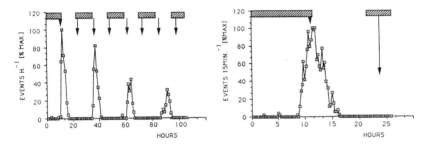

Figure 3-16 Left: A direct plot of the spontaneous locomotor activity of a single *C. maenas*. Symbols as in Fig. 3-15. Note that on the first 2 days a solar-day peak arose in the middle of the expected night. Right: A 25.75-h form estimate of the same data (Williams, 1991).

minuscule — certainly undeserving of a figure. In the conditions described in (2), with high tides in the early evening and just before dawn, 10 of 16 crabs showed a peak at the times of expected dawn high tides and in the middle of would-be night-time (when high tides are not scheduled). The latter peak appears to be an expression of the solar-day rhythm, and the former, part of the tidal one. The other six animals in this series expressed only a dawn peak (Fig. 3-16). In condition (3) animals, one third were active only in the middle of the night, i.e., there was no indication of a tidal rhythm. Seven of them were active at night and during the expected time of the morning tide. The remainder of the crabs were active only at the expected times of the morning tide (Fig. 3-17). Oh to return to the prelapsarian innocence of using only grouped data.

Needless to say, it is very hard to reconcile segments of these data with one another, and all of it with the earlier work on this animal done by the rest of Naylor's group and described above. To distill Williams' findings

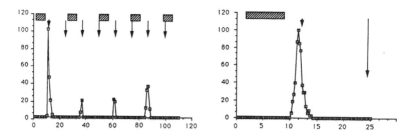

Figure 3-17   The spontaneous activity of a single *C. maenas* in constant conditions for a 4-day interval. Symbols as in Fig. 3-15. Note that peaks built up only at the expected times of *daytime* high tides (Williams, 1991).

somewhat, only a third of the animals showed any evidence of a two-peaks/day rhythm, and in these animals one of the partner peaks was often barely discernible, and often not expressed for the whole 4-day stint in the laboratory. Additionally Naylor (1958) had found that only when both tidal peaks were displayed during the daytime, was a solar-day peak expressed at night. In the Williams study, the night-time solar-day peak was often found in addition to a night-time tidal peak; and if a circadian nocturnal peak was expressed, two tidal peaks were *never* shown. These are maddening results, the kind that could drive some investigators to the anodyne of self-pithing. Thus, the Williams results have produced a conundrum that will hopefully stimulate a great deal of additional research effort into the already particularly fascinating behaviors of this animal.

The Penultimate-hour Crab (Sesarma reticulatum)

This semiterrestrial crab can live side-by-side with *Uca* and *Carcinus* in the intertidal zone. It sits out low tides in self-made burrows crowned with chimney-shaped entrances, and roams the sea floor during high water, especially at night just prior to midnight. Like the green shore crab, this decapod also contains both tidal and daily components in its locomotor pattern.

In the first laboratory study (Palmer, 1967) of its activity rhythms (Fingerman *et al.* (1961) had already described the animal's color-change rhythm), only the group responses of ten-crab samples were considered: Fig. 3-18 is a compact plot of two, ten-crab combines, a second fresh batch replacing the first on day 13. The zeros stand for hours in which activity was greater than a daily, mean-hourly. Notice the obvious clustering mainly prior to midnight (a form estimate shows the peak at 11 p.m. — thus the hypocoristic common name of the animal). In constant darkness, the rhythm is almost always indistinguishable from 24 h. The diagonal lines superimposed over the figure connect the mid-points of high tide taken from a published tide table, and clearly reveal the presence of a tidal

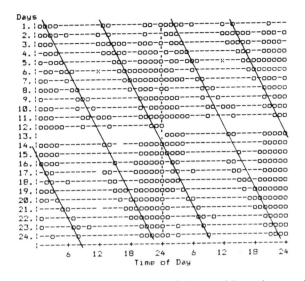

Figure 3-18 A double-plotted actogram of the penultimate-hour crab, *Sesarma reticulatum*, spontaneous activity while maintained in constant darkness at 20°C. The figure is constructed from the combined response of two groups of ten crabs each, the first group for days 1–12, and the second for days 14–24. The parallel, diagonal lines superimposed over the data connect the midpoints of high tides on consecutive days (as taken from a tide table), and thus identify the presence of the tidal component. The other prominent group of high value-points cluster around the vertical dashed line arising from 24 h, and indicate the presence of a solar-day component (Palmer, 1990a).

component. Seiple (1981) subsequently confirmed the existence of the tidal component, but did not test the persistence of the solar-day rhythm in constant conditions.

The rhythms of *Sesarma* are among the noisiest of intertidal dwellers, making the species far from ideal for study. Nevertheless, being benighted and easily amused, I embarked on a 4-year study of the *individual* rhythms of this fractious creature. The following story, outlining just how complicated the display actually is, is based on the activity patterns of 86 crabs. Twenty percent of this group showed *only* a daily rhythm (Fig. 3-19); 24% displayed *only* a tidal rhythm (Fig. 3-20), 21% expressed *both* daily and tidal components (Fig. 3-21), while the response of the remaining 35% was obnubilated beyond comprehension. The **xs** in the upper right-hand corner of Fig. 3-20, plus some of the fuzzy tails on the daily **c**-peaks, may be the remnants of a short-lived solar-day rhythm. The daily onset of the **c**-peak is much more precise than that of the **e**-peak; the latter, starting somewhere between days 9 and 14, clearly splits in two. The form estimate in the lower half of that figure is based on all the data realigned to 24.7 h; the arrow showing the presence of the fragmentation branch. Only three of the crabs provided clear examples of peak splitting.

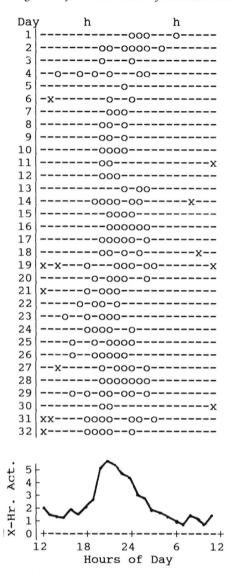

```
Day          h                    h
  1 |-------------ooo---o-----
  2 |--------oo-oooo-o-------
  3 |--------o---o-----------
  4 |--o--o-o-o---oo---------
  5 |------------o-----------
  6 |-x-------o--o-----------
  7 |---------ooo-----------
  8 |--------oo-o-----------
  9 |--------oo-o-----------
 10 |-------oooo-----------
 11 |-------oo-------------x
 12 |-------ooo-----------
 13 |------------o-oo---------
 14 |-------oooo-oo------x---
 15 |---------oooo----------
 16 |--------oooooo--------
 17 |-------ooooo-o---------
 18 |--------oo-o-o------x--
 19 |x-x---o---ooo-oo------x
 20 |-------o-ooo--o--------
 21 |x-----o-ooo-----------
 22 |-----o-oo-o-----------
 23 |---o--o-ooo-----------
 24 |------oooo--o---------
 25 |----o--o-oooo---------
 26 |----o--ooooo---------
 27 |--x-----o-ooo-o-------
 28 |-------ooooooo--------
 29 |----o-o-oo-oo-o-------
 30 |--------oo------------x
 31 |xx----oooo--oo-o-------
 32 |x-----oooo--o---------
```

```
     5 |-
X-Hr. Act.
     4 |-
     3 |-
     2 |-
     1 |-
     0 |+-----+-----+-----+-----+
        12    18    24     6    12
              Hours of Day
```

Figure 3-19 The solar-day rhythm of an individual *S. reticulatum* exposed to constant darkness at 20°C. The period of the rhythm meandered around 24 h. At the base is the form estimate for the 32-day string of data above. Note that there is no indication of a tidal component. The **h**s represent the times of the expected high tides on the first day in constant conditions (Palmer, 1990a).

Figure 3-22 is that of a crab showing only a tidal rhythm while exposed to a daily light/dark cycle. The latter does not entrain the rhythm (of course) but modifies it slightly in two ways: 1, the amount of activity is momentarily increased as a tidal peak enters light-on or light-off; 2, the period is slightly, and only temporarily, altered as it passes through the dark

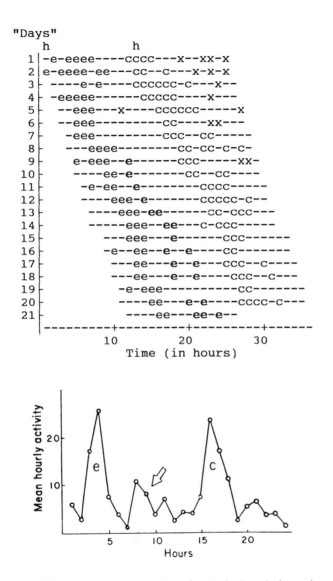

Figure 3-20 The tide-associated rhythm of a single *S. reticulatum* in constant darkness at 20°C. *Top:* Both the **e** and **c** rhythms have average periods of 24.8 h. Sometime between days 9 and 14 the **e** rhythm split in two. Note the absence of a solar-day component. Just when the story was getting interesting here, Hurricane Gloria struck and terminated electrical services for a week. *Bottom:* a form estimate for days 14–21, showing the branch (arrow) that came off from the **e** peak (Palmer, 1990a).

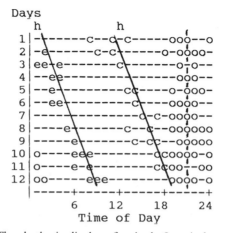

Figure 3-21   The rhythmic display of a single *S. reticulatum* in constant darkness at 20°C, showing two tidal peaks, **e** and **c**, and a solar-day component, **o**. The tidal peaks have periods of 24.9 h (Palmer, 1990a).

interval. The former (the activity bursts) might be considered "masking" (Page, 1989), and as such, disappear as soon as the light/dark cycles are withdrawn. The latter is "relative coordination" (Holst, 1939), meaning that the light/dark treatment is insufficient to actually entrain the rhythm to its frequency (Palmer, 1990a).

The Cranny Crab (Cyclograpsus lavauxi)

This decapod lives at the upper extremes of the intertidal zone throughout almost all of New Zealand. It hides under rocks during daytime low tides, but ventures forth during high tides, especially at night. When brought into the laboratory and exposed to constant darkness, 90% of the crabs displayed circalunidian activity rhythms. Like other crabs, one of the daily tidal peaks occasionally disappeared or split without the other peak being affected. Most interesting is the influence of a concomitant solar-day rhythm modifying the tide-associated one. The actogram depicted in Fig. 3-23, of a single crab in constant darkness and 14.5°C, shows the **e** peak disappearing after just 6 days in constant conditions, while the **o** peak continued unabated for 34 days, running with an average period of 26.5 h. The vertical lines arising from the abscissa at each 10 p.m. and 6 a.m. point, mark off the beginning and end of what had been night-time when this crab was captured. Note that each time the **o** peak passed through these former nights that the density of **o**s increased, signifying greater activity. The mean activity increased 187% between 10 p.m. and 6 a.m., versus the other hours of the day, and this difference was highly significant ($P = 10^{-9}$) (Palmer & Williams, 1993a).

Like the fiddler crab (Palmer, 1964a), this crab also exhibited a circadian rhythm in its response to light (Palmer & Williams, 1993a).

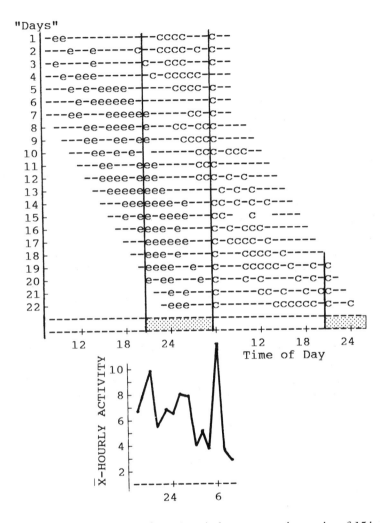

Figure 3-22 A compact plot of one *S. reticulatum* exposed to cycles of 15 h of light alternating with 9 h of darkness, and a constant temperature of 20°C. As a tidal peak passed into or out of darkness, the transition caused a burst of activity, as can be seen in the form estimate below the dark exposure (signified by the cross-hatched bar at the base). The spikes are just a response to the change in lighting and disappear when the light/dark cycle is discontinued (Palmer, 1990a).

The Pliant-pendulum (Helice crassa) *and Schizo* (Macrophthalmus hirtipes) *Crabs*

These two semiterrestrial crabs are common decapods on the sand and mud flats near the Portobello Marine Laboratory on the South Island of New Zealand. The crabs live in the intertidal zone of the Papanui Inlet, where the tides are semidiurnal and equal in amplitude. Early field work by Beer (1959) and Fielder and Jones (1978) described the animals as being most active at the times of low tide. But in 1985, Williams and colleagues set pitfall traps in the crabs' habitat and learned that the majority of their activity took place underwater during high tides. Furthermore, studying the group responses of five-animal samples for 2.5-day intervals in the laboratory, they found clear-cut persistent rhythms in both species, with peaks coming 12.4 h apart.

Using this introduction, Barbara Williams and I began a several-year study of the individual responses of these two crabs to constant conditions. The variability between specimens was large. In one sample of 28 pliant-pendulum crabs, 68% of the animals expressed the usual two peaks/lunar day cadence, while 25% displayed only one peak/day. The remainder were arrhythmic. The periods were all circa, ranging from 25.0 h to 26.9 h (mean ± standard deviation = 25.7 ± 0.5 h). The mean for the unimodal group was longer, 26.5 ± 1.4 h, with a range of 24–27.9 h. Some of the animals' rhythms were followed in the laboratory for as long at 219 days (Williams, 1991)!

The results for the schizo crab were similar. The most typical response has already been presented: Fig. 2-7 is a compact plot of a single *M. hirtipes* maintained in dim red light at 15°C, plus a form-estimate constructed by realigning all the data (not just that filtered, as in the compact plot) to 24.5 h (Palmer & Williams, 1986a, 1986b; Williams & Palmer, 1988).

A few members of both species demonstrated properties supporting the circalunidian hypothesis. In Fig. 3-24 is a clear example of a single *M. hirtipes*'s rhythm whose peaks scan the day at different rates (the difference is statistically significant: *P* <0.001). The most extreme example we encountered is shown in Fig. 3-25; the two peaks collided about day 11, and during the next few days just what happened is problematical, but for what it worth, my guess, based on the entering trajectory of **e**, is that it passed right through the **o** peak and took up residence on the far side.

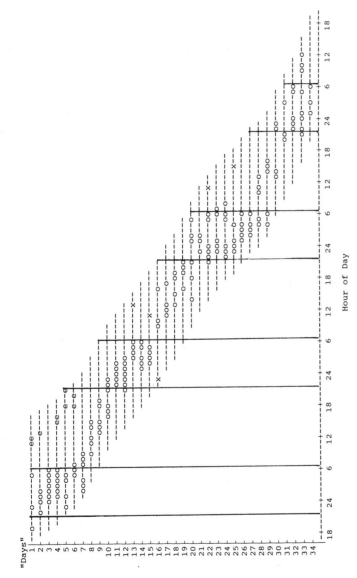

Figure 3-23 The persistent activity rhythm of the cranny crab *Cyclograpsus lavauxi*. First note that the **e** peak persisted for only 6 days. Then see that the **o** peak, which ran at an average period of 26.5 h, increased in the amount of activity displayed as it passed through the interval between 10 p.m. and 6 a.m., the hours of night-time when the crab was collected. The average 187% increase during these times differed significantly ($P = 10^{-9}$) from the subjective daylight hours. See text for more details (Palmer & Williams, 1993a).

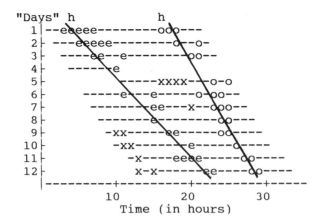

Figure 3-24 A compact plot of the spontaneous activity of a single schizo crab (*M. hirtipes*) maintained in constant dim red light and 15°C. The **e** and **o** peaks scan the day at different rates: 25.4 h for **e**, and 24.9 h for **o** (*P*<0.001). There was a recorder failure on day 4 (Palmer & Williams, 1986b).

The regression lines were fit by eye, and then used to realign the separate peaks so that form estimates could be constructed. Considering the sharpness of the peaks of these estimates, and their small standard errors, suggests that my guess (above) may really be correct. (Thinking like a crab helps also.) A significant (*P*<0.005) difference in the scan rates of the peaks of *Helice* crabs was also found (Fig. 3-26) in a few animals (Palmer & Williams, 1986b).

In these two species, just as in those previously described, one or the other of the two peaks sometimes split into two, or more rarely three, different fragments, while the other remained unchanged. In fact, it was in these two crabs that splitting was first seen in tide- associated rhythms (Palmer & Williams, 1987), although it is quite common in circadian activity rhythms, especially in small mammals.

Using MESA and autocorrelation, short-period oscillations — ultradian rhythms — have been found to be a part of the displays of these two crabs (Dowse & Palmer, 1992), just as they were in *Uca*. One example should suffice: Fig. 3-27 is a MESA plot of a single *M. hirtipes* showing the typical *circa* tide-associated peaks at 13 h and 25 h, and a very strong spike at 8.5 h. The ultradian interval thus identified, an actogram plotted modulo 8.5 h (and duplicated in the usual manner) was made, and is presented in Fig. 3-28. Here one sees that by day 9 the ultradian period is not only identifiable, it is dominant. Its period holds fairly constant through day 29, and then begins to wander back and forth around that value up to day 89.

As just described, it is clear that these ultradian rhythms are robust and long-lived, and are not merely subtle variations at the edge of detectability.

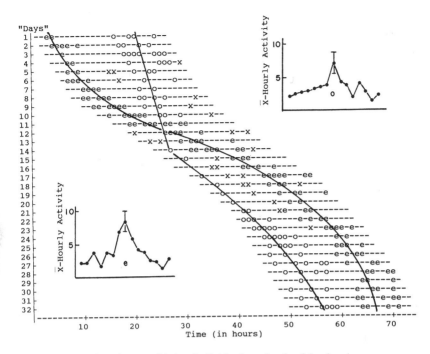

Figure 3-25   Another *M. hirtipes* individual, maintained in the same conditions as the animal in Fig. 3-24. The interpretation here (a conclusion not etched in stone) is that the **e** and **o** peaks assumed such different periods that they actually crossed! The regression lines, fit by eye, were used to realign the data of each peak and construct form estimates of each (insets). The sharpness of the peaks thus described, plus the gradual way the values build up to the maximum and fall off thereafter, suggests that the above supposition may actually be somewhere near reality (Palmer & Williams, 1986b).

Thus, what is their significance? There can be no ecologically defensible reason for their existence: they have no relationship to any geophysical cycle known; they are not submultiples of the tidal interval; nor do their period lengths fall around some average — instead they are distributed evenly throughout the spectrum. Are they just an epiphenomenon? When the circalunidian periods weaken the ultradian often become more prominent and vice versa. Their existence provides a challenging conundrum that is discussed further in Chapter 6.

Peracarids

The intertidal burrowing amphipod, *Corophium volutator* (Figure 6-17, p. 190), sits out low tides in its burrow, emerges with the flood tide to be swept shorewards, and swims maximally during ebb tide. It is thought that maximum swimming at ebb tide helps guarantee that the animals do not get stranded on dry land as the tide recedes. When their swimming activity

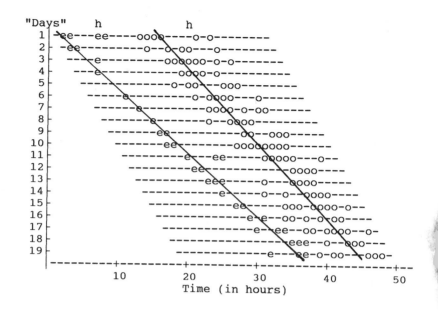

Figure 3-26 Compact plot of a single *Helice* crab exposed to constant dim red light and 15°C. Peak **e** has an average period of 25.8 h, and peak **o** one of 25.6 h (*P*<0.005) (Palmer & Williams, 1986b).

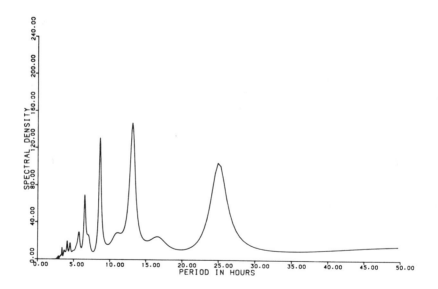

Figure 3-27 A MESA plot showing significant periods at 8.5 h (ultradian), 13 h, and 25 h (*circa* tide-associated) periods (Dowse & Palmer, 1992).

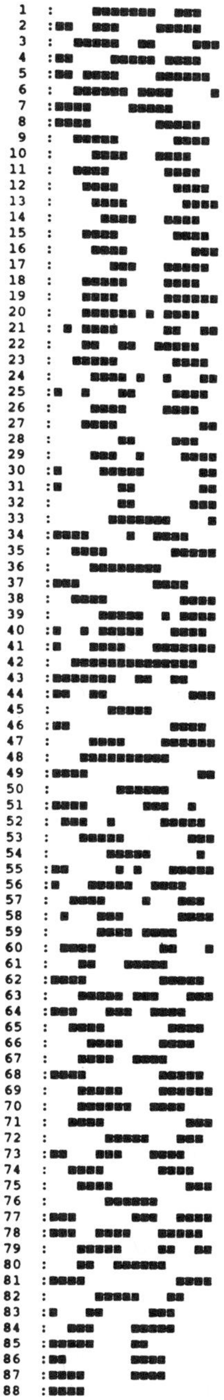

Figure 3-28 The same data analyzed in Fig. 3-27 realigned modulo 8.5 h. To increase viewing ease, the actogram has been duplicated and the copy appended to the right of the original. Starting with day 9 and ending on about day 29, the 8.5-h period remained fairly stable. After that day it wandered back and forth around that interval (Dowse & Palmer, 1992).

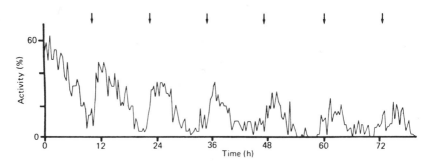

Figure 3-29   The swimming rhythm of ten amphipods (*Corophium volutator*) in constant conditions. The rain of arrows fall at the times of expected high tides (Holmström & Morgan, 1983).

is studied in constant conditions in the lab using time-lapse photography, this rhythm has been found to persist for 3–6 days (Fig. 3-29). By replacing laboratory animals with freshly collected ones over fortnightly intervals, very slight changes in amplitude, phase, and period length were found between spring and neap tide swimming patterns.

Animals collected from essentially non-tidal pools (those highest on the shoreline and thus reached only by the greatest spring tides) are found to display neither tidal nor solar-daily rhythms in the laboratory.

Storage in the laboratory rather quickly causes arrhythmicity, and restoring this loss has taught investigators a great deal about just what facets of the tides function as phase setters and entraining agents. I will whet your appetite here for what is to come in the next chapter. When arrhythmic amphipods are placed in trays covered by plankton netting and set out on the shoreline, exposure to as few as six tidal exchanges causes the rhythm to be reinstilled. But if the container of amphipods is floated instead just below a buoy on the surface in calm weather, there is no rhythm startup (Morgan, 1965; Holmström & Morgan, 1983; Harris & Morgan, 1986).

The animal also undergoes a persistent rhythm in oxygen consumption. As would be expected, maximum oxygen uptake is in exact phase with maximum activity, obviously a cause-and-effect relationship (Harris & Morgan, 1984).

The predaceous isopod *Eurydice pulchra* has a life-style almost identical to *Corophium*: it lives buried in the intertidal sand during low tides but rises into the water column with the flood to swim by the water's edge, feeding on infauna and debris, as the tide moves farther inland. It then retreats seaward with the ebb and reburies itself (Warman *et al.*, 1991). The swimming rhythm persists in constant conditions in the laboratory, with peak swimming occurring at the times of expected ebb tide. The night-time peak is of greater amplitude, and there is also a fortnightly modulation in the amplitude of the activity: it is greatest during the spring tides (Jones & Naylor, 1970; Fish & Fish, 1972; Hastings, 1981a; Reid & Naylor, 1985).

In a different experimental design, the isopods were maintained immobilized in wet, sterile sand and their oxygen consumptions measured. The animals had been collected during the spring tides and were maintained in constant darkness and at 15°C. A strong rhythm was found phased to the times of expected high tides; its peak O_2 consumption was 50% greater than the basal rate. Because the animals were immobilized at the time of recording, the rhythm could not have been just a secondary consequence of swimming movements (Hastings, 1981b).

The New Zealand Clockle *(Austrovenus stutchburyi)*

Intertidal bivalves are periodically exposed to air, during which time most keep their shells tightly closed; during immersion they open up so as to filter feed. In 1954, Rao reported shell-gaping rhythms in two species of mussels (*Mytilus californianus* and *M. edulis*) that persisted more than four weeks in the laboratory. Curiously, their rhythms ran at exact 12.4-h periods the whole time and, unlike Naylor's finding with green-crab rhythms, even mussels attached to a floating raft described the same precise tidal rhythms. Additionally, those members living subtidally also had persistent tidal rhythms! How unusual; and investigators in other laboratories (Jorgensen, 1960; Davids, 1964; Thompson & Bayne, 1972) have been unable to confirm Rao's findings.

Several workers have described shell-gaping rhythms in the quahog *Mercenaria mercinaria*, and the oyster *Crassostrea virginica* (Bennett, 1954; Brown, 1954; Brown *et al.*, 1956a, 1956b); but Enright (1965) using a more modern statistical approach has questioned the reality of those rhythms. Palmer (1980) was unable to find a filtration-rate rhythm in *C. virginica*. Morton (1971) conducted a one-day study that revealed a tide-associated rhythm in crystalline-style volume in *Ostrea edulis*; a feature that might be expected to accompany the feeding cycle associated with shell gaping.

The first indubitable study was done by Beentjes and Williams (1986) on the New Zealand clockle *Austrovenus stutchburyi*, a delicious-tasting bivalve, with a portmanteau-word name, living intertidally near the Portobello Marine Laboratory in New Zealand. Sample populations of 36–64 of these animals were placed on their sides in a sea table supplied with either running water, or standing-aerated water. The light was held constant and the number of clams with open shells counted each hour by a faithful student. While such an experimental design seems unusually unsophisticated in this era of high-tech electronic biology, it had two important advantages: (1) it worked; (2) to keep it running reliably, the only maintenance required was an occasional meal for the student. The results are seen in Fig. 3-30. In further studies threads were attached to

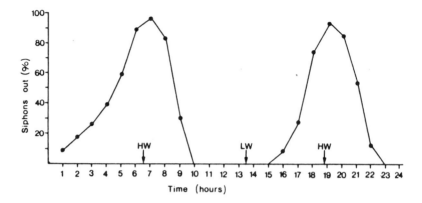

Figure 3-30 The percentage of 48 *Austrovenus stutchburyi* clockles with shells agape while kept in constant illumination and a relatively constant temperature. HW and LW stand for high and low water (Beentjes & Williams, 1986).

the valves and the movements recorded automatically and continuously (except for breakdowns) on a kymograph; the shell-gaping rhythm was found to persist for as long as 2 weeks. It is interesting to note that even though the clockles were at all times submerged, they still closed their shells during the "clammed up" phase of their rhythm. For those animals maintained in running water, who were thus constantly surrounded by a food-laden environment, it meant passing up easy chances to double their nutrient-gathering opportunities.

This introductory work was followed up by a study of individual shell-gaping behavior. One valve of each clockle was glued with dental cement to upright underwater partitions, with their valve margins facing upwards and hinges down. A video camera was mounted overhead and turned on for 30 s every half hour, recording which *Austrovenus* were open, and which ones had closed. The subjects were maintained at 14.5°C in constant light, in non-circulating aerated water. As usually happens when individual rhythms are studied, some unanticipated results were obtained. The basic pattern is seen in Fig. 3-31. The rhythms are unique in that many peaks are skipped. In one study of 30 individuals there was a one-in-three chance that a peak would be expressed on any one day. In an extreme case, no peak (i.e., the valves remained closed) was manifested for 11 straight days! No other tide-associated rhythm studied thus far has been found to be so sporadic in its expression, and there is no explanation for this occasional silence of the clam (Palmer & Williams, 1993b; Williams *et al.*, 1993).

When a peak is finally aired after several days' absence, however, it falls on an invisible trend line extrapolating from pre-expressed peaks. Our interpretation is that the clocks underlying these tide-associated rhythms are quite accurate (meaning that *clockle* is an apt sobriquet), but the coupling mechanism between clock and shell gape is fickle (*vagarious-*

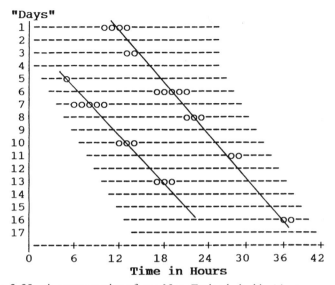

Figure 3-31 A compact plot of one New Zealand clockle (*Austrovenus stutchburyi*) in constant light at 14.5°C. The **os** (= open) indicate the hours of shell gape. Although peaks are most often not manifest, it is clear that when they do arise they come at a pre-ordained time. The average periods here are *c*.25.7 h (Palmer & Williams, 1993b).

coupler clam would be equally appropriate). Why such an incomparable combination should exist is curious and incongruous, and thus hard to explain. One might speculate that natural selection increased the clock's precision as a way to compensate for the inadequate coupler (that may seem ludicrous, but nature, as a mother, does it her way: look at the design of the human menstrual cycle). Or, more likely, the clock was designed to govern some other process than shell gaping, e.g., some physiological endeavor such as crystalline-style synthesis (an event known to be rhythmic in this species [Hutchinson, 1988]) that must be initiated in *anticipation* of the returning tide. Such a preparatory action would be temporally more important to a clockle because, one would think, the animal can simply open its shell anytime to test for changing events in the tidal environment.

The average period length in these conditions for 61 animals was 25.7 ± 0.43 h (range = 24.8–26.6 h).

By this time, a reader should not be surprised when I report that a great variability was found between the rhythms of different individuals. I will parade a few samples for viewing. The first (Fig. 3-32) shows that the period in the laboratory can change spontaneously, and that one peak may disappear at times. In this case the period changed from 26.1 h to an uncommonly short 22 h.

The next pattern is another extreme example (Fig. 3-33). Not shown

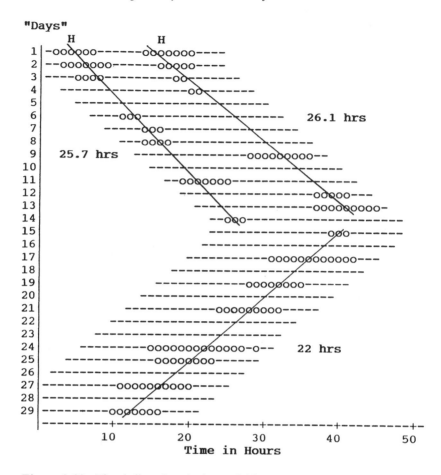

Figure 3-32 The shell-gaping rhythms of this *Austrovenus* assumed different periods (*P*<0.001) during the first 14 days. Then one peak vanished, while the other shortened unusually to a length of 22 h. H = the expected time of high tide on day 1 (Palmer & Williams, 1993b).

is the fact that this clockle (who we nicknamed Buridan's ass) did not open its shells for the first 5¼ days in constant conditions, then held them open for the next 26 h, then clammed up for 58 h, opened briefly, closed for 1½ days, and finally settled on a unimodal display with a stretched period length of *c*.33 h!

Figure 3-34 describes an especially clear-cut tide-associated response that suddenly became a unimodal expression. One can only guess what happened, but one interpretation is that the two peaks spontaneously fused. The unimodal period first shortened to just under 24 h, and then lengthened to *c*.26 h.

In this species also, we find the occasional animal in the laboratory that

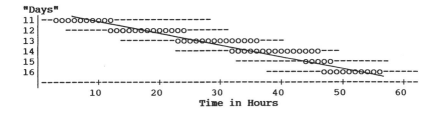

Figure 3-33 This *A. stutchburyi* did not become overtly rhythmic until day 11, and then produced only a one-peak display with an exaggerated period of *c*.33 h (Palmer & Williams, 1993).

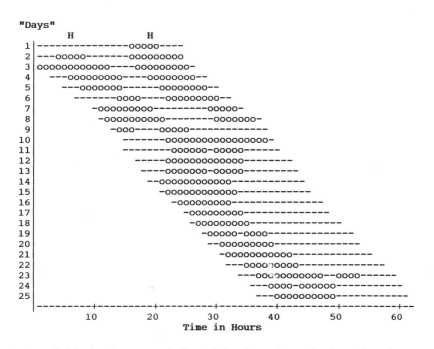

Figure 3-34 An *Austrovenus* individual that began its rhythmic display with a non-traditional clam show — a near-perfect performance. But on day 10, it appears that the two rhythms fused, shortened to just under 24 h, and then lengthened again to *c*.26 h (Palmer & Williams, 1993b).

produces peaks that scan the solar day at different rates. The rhythms displayed in Fig. 3-35 have periods of 25.7 h and 26.0 h in length. The 15+ min difference is significant: $P = <0.02$. Notice also the period differences in the early display of the clockle pictured in Fig. 3-32; the divergence here is also significant: $P<0.001$.

Here, then, is yet another species whose individual responses to

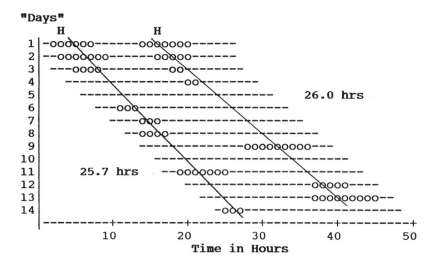

Figure 3-35 A compact plot of a single *Austrovenus* showing period lengths that differ significantly (*P*<0.02) from one another (Palmer & Williams, 1993b).

constant conditions fully support the reality of the circalunidian hypothesis of rhythm control.

The Basket Cockle (Clinocardium nuttalli) as a Geochronometer

There is a branch of paleontology that is concerned with the rate at which the earth's rotation is slowing down: in the mid-Devonian, a mere 200 million years ago, days were only *c*.22 h long! Claims such as that are based on counting the layers in fossilized coral thecae, stromatolites, and mollusk shells: one layer is thought to be added each day, and because the layers are thicker in one season than another, intervals of a year can be identified. Thus, it is claimed that back in the Devonian there were 400 days annually (Wells, 1963). Thus, fossil corals have been considered geochronometers or paleontological clocks because of this banding. To be sure, one would have had to be there to verify the deceleration, but hindsight evidence is pretty convincing that on average the earth is losing about 2 s every 100,000 years (ah my, where does the time go!).

Modern marine chronobiologists have benefited from these kinds of paleontological endeavors, because as a first step in surmising what ancient organisms did, those investigators have to see what extant ocean dwellers are doing now. Some animals, especially the corals (probably because of the zooanthellae they contain) often do produce one thecal layer/day. But intertidal clams, especially those that live just under the upper layers of soft sand (where they experience the full force of the tides), often add a layer of calcium carbonate ($CaCO_3$) in an organic matrix (called conchiolin) on the inner surface of their valves during each high tide when they gap to filter feed (meaning that an uninformed paleontologist, using the one layer/day "rule," would decide that our years had about 707 days, and philosophers would mourn that our life span was only 36 years! — but we would have two summers each year). During the time that the valves are open and the clams are filter feeding, aerobic conditions prevail inside the mantle cavity. When the shells are closed, the mantle cavity becomes anaerobic and organic acids build up, causing some calcium carbonate to be etched away leaving the darkish conchiolin prominent (Lutz & Rhoads, 1977). The ritual engraves the interval of a tidal cycle into the shell.

The basket cockle *Clinocardium nuttalli*, is a sand-dwelling bivalve living in the intertidal on the west coast of the United States, where, as mentioned before, the tides are mixed, semidiurnal, and unequal. When acetate peels are made of a sectioned shell, micro-growth lines are easily distinguished and counted. A very convincing photograph of this tide-associated rhythm in shell deposition is offered in Fig. 3-36. In the bottom half of this figure is seen the thin section of the shell: the dark lines (except for the jagged crack) are bands of conchiolin, while the light bands, varying in thickness, are inorganic shell material. The waveform drawn above the shell represents the tides to which the animal had been exposed; the straight horizontal line skewering it represents the position on the shoreline where the clam had lived before it began to help the advancement of science. All of the lows in the waveform are, of course, intervals of low tide; note that the pointer-lines projecting downward from these troughs lead to conchiolin bands — entities formed when the valves were held closed. What we are seeing here is a historical record of tidal exchange on the Oregon coast.

Fig. 3-37 presents a 15-month shell-growth record of the same cockle presented as a histogram: the length of each bar indicates the width of the growth increment (Evans, 1975).

The tide-associated deposition of shell material in the prismatic layer of the intertidal mussel *Cerastoderma edule* has also been demonstrated. The rhythm is prominent in freshly collected animals, and persists (although the bands are much fainter) in the atidal laboratory (Richardson *et al.*, 1980).

Were it not such a labor-intensive chore, it would be interesting to carry out the same kind of study on the New Zealand clockle, *Austrovenus*, since its persistent shell-gaping rhythm is now well documented (above). It would tell us if the clam sometimes remained closed for days at a time in the field, just as it does in the constancy of the laboratory.

Fishes

There are two general types of behavior shown by the fishes that inhabit the intertidal zone. Rocky-shore species sit out intervals of low tide

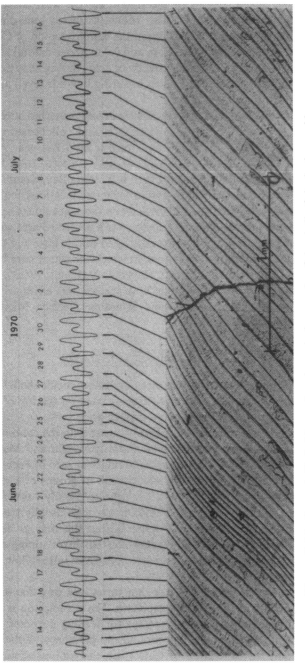

Figure 3-36 *Lower half:* A thin section through the margin of the shell of the basket cockle *Clinocardium nuttalli*, showing alternating layers of calcium (light bands) and conchiolin (dark bands). *Upper half:* a representation of the tides to which this bivalve had been exposed. The horizontal line superimposed over the tidal schedule represents the level on the shoreline where the clam had lived. Using the pointer lines falling from each tidal trough you can see that during each interval of low tide a strip of organic conchiolin was exposed (Evans, 1975).

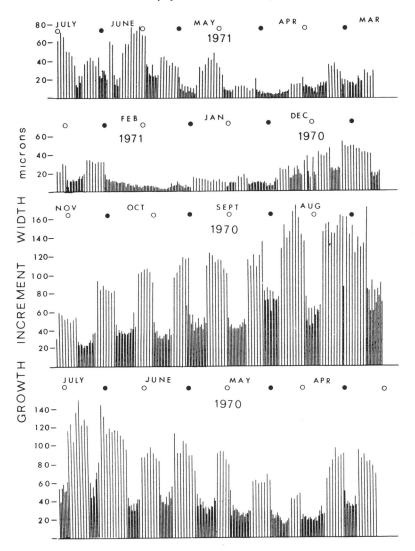

Figure 3-37   A 15-month record of growth in the same cockle shell seen in Fig. 3-36. The height of each bar represents the width of one layer of $CaCO_3$; the widely spaced bars describe layers deposited once every lunar day (during spring tides), and the narrow bars describe depositions made twice every lunar day (at neap tides). The fortnightly alternation in layer thickness is obvious. The solid and open circles represent new and full moons, respectively. Note that locked in this shell is a 1.5-year history of past tides (Evans, 1975).

"stranded" in tide pools, while sandy-beach species swim inshore with the flooding tide and retreat back down the intertidal with the ebbing tide. In the former, maximum activity is locked to the time of high tide, while in the latter it is phased to the ebb. That relationship in the sandy-beach

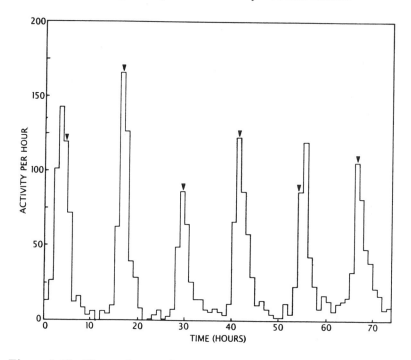

Figure 3-38 The persistent swimming activity of the shanny, *Lipophry pholis,* for six cycles. The rhythm persists in either constant light or darkness (modified from Gibson, 1965).

species is thought to be adaptive, insuring that the fish do not become stranded on dry land.

The tidal activity pattern of many fish species in known to persist in the constancy of the laboratory (Gibson, 1982, 1992). In several ways their rhythms are similar to those of crabs, the great variability between individuals being a common trait. As Gibson (1976) points out, in his first experiments only about 10% of the subjects described persistent rhythms precise enough to be useful. Thus, often the responses of several individuals were grouped in an attempt to portray a "typical" response. Needless to say, that led to a loss in definition in the display. Another problem was that the rhythms damp out after a few days.

Figure 3-38 is a superb example of a persistent rhythm of a shanny, *Lipophrys pholis* (Gibson, 1965). This animal's rhythm is so precise it probably deserves a "supershanny" appellation. In a large, more recent study it was found that the shanny swimming rhythm persisted in 69% of the subjects for at least 6 cycles, but damps considerably and loses precision during these days. The mean period length ($\pm$ one standard error) for the group was approximately 12.89 ± 0.66 h, while the tides in the animals' habitat averaged 12.4 h (Northcott *et al.*, 1990).

Like crab rhythms, occasionally a peak will split, but unlike crabs it usually fuses again quickly. Also like crabs (and clams) one of the two daily peaks will disappear and return again later. In one fish only a single peak was displayed, and it ran at a period of 25.5 h. The maxima of this unimodal rhythm were large, and the significance of the period high (by periodogram analysis: $P < 0.01$). Could this have been the expression of a single circalunidian clock?

Commonly living with *L. pholis*, often cohabiting the same rock pools, is the goby *Gobius paganellus*. These two species, along with the very rare *Coryphoblennius galerita*, are the only truly intertidal species in the United Kingdom. As with *L. pholis*, *G. paganellus* displays a persistent tide-associated rhythm in the laboratory, but, unlike the former, its waveforms are not as precise. Additionally, the phase of every other peak alternates in its relationship with expected tides: one leads only slightly and the other leads more so and the lead increases with time in the lab. This is just another way of saying that the periods of the two peaks differ, although in this case not significantly so. If, when more studies are undertaken (this one included only 16 fish), it should be found that this is a common feature of the goby rhythm, it will add more support to the circalunidian hypothesis (Northcott, 1991).

The fact that *some* of these individuals fish have the most clear-cut rhythms of their intertidal finned companions, and that their rhythms are short-lived in constant conditions, has made them ideal subjects to exploit in experiments designed to identify factors in the environment that can re-establish lost rhythmicity and entrain basic tidal-rhythm ability to the tidal schedule of a particular coastline. Drs Sally Northcott, Robin Gibson, and Elfed Morgan have capitalized on these properties, as will be described in the next chapter.

Are There Moon-Related Rhythms in Humans?

Scattered through popular literature, poetry ("moon-spoon-June"), and song are suggestions and claims that the moon overhead has an influential effect on human life — especially on our sex lives (including sex crimes!). In my first, and last, (ad)venture into the social sciences, I teamed up with Dr Richard Udry at the Carolina Population Center in a test of such supposition. Seventy-eight couples were paid $1 each day to mail in a postcard reporting if, and at what time, they had had intercourse in the previous 24 h. In this way we collected 5,584, of what I will call "copulating-couple days" of data.

A most obvious daily copulatory rhythm was present in the data. As you can see in Fig. 3-39, the greatest number of copulations (56%) occurred between 2100 h and 0100 h. There is a secondary hump at 0700 h which happens to be the time of day that the testosterone level of blood is greatest in males (Resko & Eik-Nes, 1966; Southern *et al.*, 1967). As an aside, a significant increase in organismic success was attained between noon and midnight.

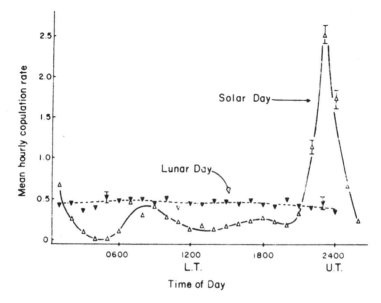

Figure 3-39 The mean solar- and lunar-day patterns of copulation based on 5,584 "copulating-couple days" of data. Some of the means have been armed with plus-or-minus-one-standard-error skewers, and those not shown on other means are all smaller. U.T. and L.T. stand for upper and lower transit of the moon. There is a prominent solar-day rhythm present, but not the slightest indication of a lunar-day cycle (Palmer *et al.*, 1982).

The data were then searched for a lunar influence. Using a method described in Palmer (1967), 30-day units of data were realigned from a solar-day arrangement into a lunar-day construction and a mean form estimate calculated. The technique randomizes the solar-day component making it vanish, and emphasizes any lunar-day residual. For instance, if the moon directly overhead had somehow tickled our libidos, a form-estimate curve with a period of 24.8 h and a maximum under the time of lunar zenith would be formed. Returning again to Fig. 3-39 you see how well the solar day pattern has been eliminated, and that there is absolutely no hint of a lunar-day rhythm. MESA and periodogram analysis both gave a peak only at 24 h.

The data were then examined in different ways. The number of copulations that took place when the moon was above the horizon (i.e., the interval between moonrise and moonset when "moonbeams" would not be blocked by the bulk of the earth) were compared to the number when it was below the horizon. The rate was 7.6% greater during the former, but the difference was not significant ($P = 0.2$). Then, because the greatest number of copulations took place between 2100 h and 0100 h, that rate was compared to moon above and moon below the horizon. Again the rate was greater with the moon above the horizon, but only a meager 1.6% ($P = 0.9$). Because Sunday is the one day of the week when a relatively undisturbed libido pattern could be expressed, one might be able to see some influence of the moon then. Thus the copulatory rates were compared with the moon above and below the horizon on the 26 Sundays in the study. The rate was 16.8% higher in the former, but the difference was again not significant ($P = 0.15$) (the form of the rhythm for that day indicates that few of the subjects went to church or the golf course; and that an alternative way had been found to fill the otherwise wasteland of a Sunday afternoon). In one last desperate attempt the number of copulations taking place during the 5 days around the cardinal phases of the moon was compared. The rate around the days of full moon was 14.6% greater, but as before, not significant ($P = 0.1$) (Palmer *et al.*, 1982).

The only conclusion that can be drawn from this study, although it was based on 1,941 copulations, is that the moon has no effect on our libidos. However, if the differences described above were to remain constant as the sample size was increased, significance would be reached. Thus, a larger sample might be tried because in all cases the copulation rate was greater when the moon was above the horizon, but that study is for someone else to do: I have enough challenge working with crabs and clams. So, at least for the time being we must leave the subject with the conclusion that there is no man in the moon (or man and woman — for those who see in that construction social benefits sufficiently great to outweigh its awkwardness), and return to the intertidal zone.

Miscellaneous Cases

The Commuter Diatom, Hantzschia virgata

This story is an old one but will be repeated here because until just recently it is the only example of a single-celled organism displaying a tide-associated rhythm.

It is not particularly uncommon to find large populations of unicellular algae living on and in the sediments of the intertidal zone. They can be so abundant that their pigments color large expanses of sand or mud, and in some places their numbers are so great that one naturalist (Herdman, 1924) reported, "... the diatoms were so abundant on the surface that their photosynthetic activity was distinctly audible as a gentle sizzling ... and the sand was frothy with bubbles of gas, presumably oxygen given off by them." Even a "lablubber" like myself can find these algae in the field! Many of the motile members of such a community remain buried in the sediments during high tides and at night, but emerge up onto the surface during daytime low tides. For a review of these *shortest* vertical migrations known to take place in the ocean, see Palmer (1976). In the great majority of cases, while the rhythm is clearly associated with the tides, when the algae are brought into the laboratory, if the rhythms persist at all, they are

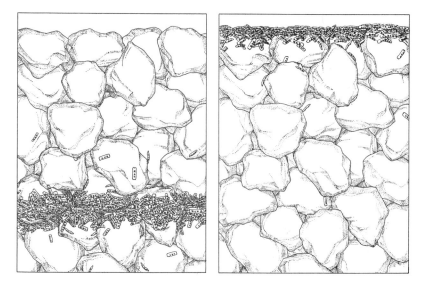

Figure 3-40 Artist's conception of the vertical migration rhythm of the commuter diatom, *Hantzschia virgata*. At the left the diatoms are seen layered on sand grains less than a millimeter below the surface; it is here they reside throughout the night-time and during high tides. The alga is motile, propelling itself upward to the surface (right-hand picture) by forcing mucus out through pores at the end of their elongated glassy cell walls. The rhythmic commutations persist in constant conditions for at least 11 days (Palmer, 1975).

usually found to be fundamentally circadian (Palmer & Round, 1965; Round & Palmer, 1966).

For years, the exception was the commuter diatom, *H. virgata*, that lives just 24 miles from the Marine Biological Laboratory in Woods Hole, Massachusetts, USA. It has a bona fide tidal rhythm. These motile cells burrow to a depth of about 200 μ and remain there until a daytime low tide exposes their sandflat; then they quickly emerge up onto the sand

---

Figure 3-41 The laboratory expression of the vertical-migration rhythm of the diatom, *Hantzschia virgata*, in (A) constant conditions (110 ft.can.; 18°C) over an 11-day interval, and (B) in a light/dark cycle for 8 days. Consecutive days are plotted one beneath the other; *x* indicates the moment of collection of the cells in the field; the wavy lines denote the times of high tide on the days of collection; the dotted lines with the dip downward denote the times of expected low tides; and the stippling represents the hours of darkness either in or out of the laboratory. Note that the peaks of the cells' rhythm follow the times of expected low tide in both laboratory situations. Because the means of studying this rhythm required removing cells from the sample population, collection of participants was skipped deliberately on some days (Palmer & Round, 1967).

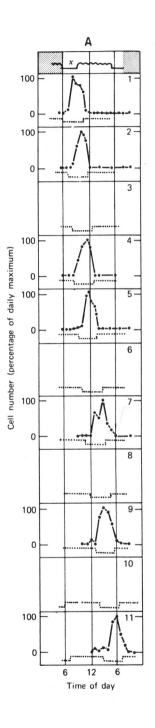

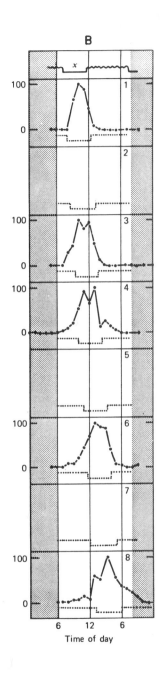

surface and pile about five cells deep, covering the underlying grey sand of that area with a rich golden-brown blanket (Fig. 3-40). It would seem axiomatic that this behavior pattern had been designed to expose the alga to sunlight for maximum photosynthesis during low tides, followed by burrowing to keep them in place during inundations. It is thus surprising to learn that this alga is actually a "shade species" because full sunlight inhibits photosynthesis by as much as 14%, and optimal light intensity penetrates more than 2 mm into the sand (Taylor & Palmer, 1963; Taylor, 1964).

During the long days of summer when at least parts of two intervals of low tide occur during the hours of daylight, the cells appear on the surface each exposure; but they never rise during night-time low tides. Light is clearly the key to this: when cells are on the surface and are experimentally thrown into darkness by placing an opaque canister over them, they burrow; and when the canister is removed, they emerge again. Because they burrow when darkened means they must be positively geotactic. It is postulated that they are always in a positively geotactic state, but when they become positively phototactic this "drive" overrides geotactic pull. Just before the tide returns they lose their photopositivity and appear to dissolve into the sand. They remain indifferent to light all through the night, as is demonstrated in the laboratory when they are exposed to constant light. This vertical migration rhythm will easily persist in constant conditions for as long as 11 days (Palmer & Round, 1967), where, as seen in Fig. 3-41A, the peaks are quite regular. The uniformity is probably a consequence of the selection pressure to which the alga is subjected: any cells that do not submerge in advance of the flood tide are washed away, leaving only the conformers to be studied in the laboratory.

Considering that most other vertical migration rhythms turn out to be circadian in the laboratory, it was possible that what we were seeing in *Hantzschia* was a circadian frequency that, by chance, happened to have a 24.8-h period. That could easily be tested by repeating the observations in a light/dark cycle: if the rhythm was circadian it would be entrained to a strict 24-h period; if it was a tidal rhythm it would not entrain. When the test was made, the rhythm was found to ignore the light/dark cycle (Figure 3-41B).

In the natural setting *Hantzschia*'s suprasurface phase comes roughly 50 min later each day until the low-tide interval begins to straddle dusk and sunset. The peak then breaks down. But the second interval of low tide, the one that has been progressing along through the hours of darkness, now enters into the morning light and the diatom's suprasurface phase appears there with its arrival, and then works its way across the hours of daylight again. We duplicated the day/night conditions in the laboratory (Fig. 3-42A). The results certainly look like the onset of darkness initiates the disintegration of the peak and a rephase to the time of light on. At first we explained this by our typewriter hypothesis: it likened the movement

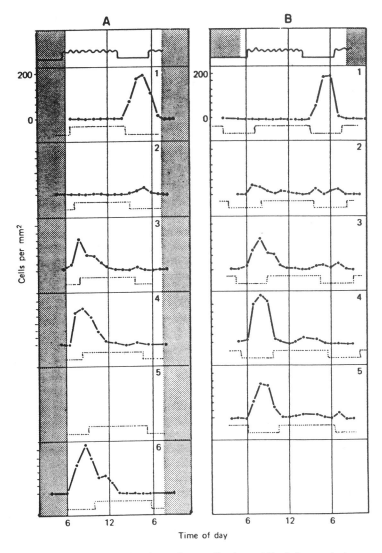

Figure 3-42 The "radical" phase-change "backwards" of the vertical migration rhythm of the diatom *Hantzschia virgata* in a light/dark cycle (A), and in constant light (110 ft.can) (B). The symbols are the same as used in Fig. 3-36 (Palmer & Round, 1967).

of the peak to a typewriter carriage that travels systematically, one character at a time, across a typewriter, and was then, with one fell swoop, swept from the far carriage stop to the near one, ready for the next trip back across the typewriter. The near and far carriage stops were light on and light off. If this hypothesis was correct, this would be the most radical phase change ever seen for a rhythm. But the half-life of this idea turned out to be just days, for the rhythm acted exactly the same in constant conditions (Fig.

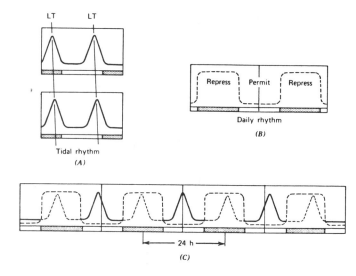

Figure 3-43 A diagram of a solar-day/lunar-day clock combination in the derivation of the overt display of the *H. virgata* vertical migration rhythm. (A) shows the tidal rhythm; (B) the daily rhythm; and (C) the combined action of the two together. LT = low tide. Stippled bars either the hours of darkness, or expected darkness. See text for full explanation (Palmer, 1974).

3-42B): carriage stops were not required! The apparent rephase appeared to be built into the rhythm!

The replacement hypothesis made no use of office-machinery surrogates; instead, two clocks were proposed, one a solar-day clock, and the other a tidal one. Because the night-time low-tide peak is not expressed even when the cells are artificially illuminated, it appears that the peak has been *actively* inhibited. The solar-day clock is assigned this function. The rhythm it drives consists of a repression phase, entrained to the night-time, alternating with a permissive phase, tuned to the daytime, which allows the expression of the tidal clock (Fig. 3-43B). Thus, a typical tidal rhythm with two peaks every lunar day (Fig. 3-43A) is transmogrified into an apparent, unimodal, lunidian overt display (Fig. 3-43C). If this idea actually represents reality, the clockworks jammed into these tiny cells is extensive and quite busy! This idea, which is only speculative, at least fits well with what is known for several other organisms (such as *Carcinus* and *Sesarma*) where solar-day and lunar-day clocks interact (Palmer & Round, 1967).

This scheme may be even more convincing when invoking the circalunidian-clock hypothesis: as each output of the lunidian clocks passes through the prohibitive phase of the 24-h rhythm, the latter would simply cut the coupling between that lunidian clock and the overt behavior it causes to be rhythmic. The strong connection between the two lunidian clocks would keep them separated at a tidal interval so as one clock emerged from the prohibitive phase, the other would enter it. The end result of this would be that the casual observer would see what appears to be a rapid rephase backwards from the late afternoon phase to a morning position.

The interest in tidal rhythms being what it is, two decades passed until another enthusiastic laboratory tackled the same rhythm. Happey-Wood and Jones (1988) studied the locomotory behavior of the diatom *Pleurosigma angulatum*, which inhabits the intertidal mud of the Menai Straits in North Wales. Field studies demonstrated that the alga underwent a vertical-migration rhythm in which cells appeared on the sediment surface only during daytime low tides. Moved into the laboratory, the rhythm was found to persist for at least 8 days under constant illumination and temperature. Like *Hantzschia*, in constant light, when the suprasurface phase of the rhythm approached what would be the expected time of sunset, a second peak spontaneously built up at the time of what would be the early morning exposure to low tide in nature. How nice to have a confirmation — albeit with another species — of one's earlier work.

An Intertidal Beetle, Thalassotrechus barbarae

This carabid is a terrestrial insect, that is a member of the intertidal "crevice fauna." At night, during low tides, it emerges from its fissure at dusk to forage over a broad area extending from about a meter above mean spring low water up to the splash zone, feeding on stranded zooplankton and herbivorous diptera larvae. Thus, in this species, we have just the opposite phase of the *Hantzschia* tide-associated, vertical-migration rhythm where peaks in the rhythm occur only in the daytime.

The beetle was collected near Stanford University's Hopkins Marine Station in California, USA, and placed, individually, in actographs maintained in constant conditions (0.05 lux and 15.5°C.) at that laboratory. The insect's spontaneous activity was measured for 5–10 days. The dominant rhythmic component — and the one verifiable by periodogram analysis — was the circadian one. In these conditions it lasted for 7 days and exhibited a period of approximately 23.9 h (Fig. 3-44). It is claimed that a tidal component existed but only for approximately 3 days; and because of its ragged form and short duration, it could not be verified by periodogram analysis. However, another form of statistical voodoo assured its reality. If this is correct, this is the only account of a terrestrial animal that has a persistent tide-associated activity rhythm (Evans, 1976).

The Cave Cricket, Ceuthophilus maculatos

This animal, also known as the camel cricket because of its humped-up posture, is really a wingless, long-horned grasshopper. Caves are, however,

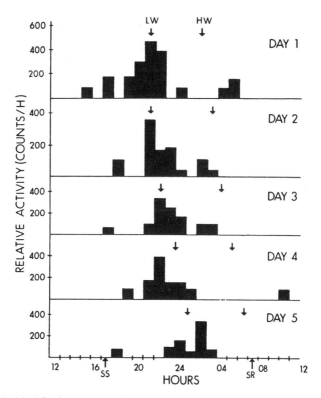

Figure 3-44 The locomotory rhythm of the intertidal beetle *Thalassotrechus barbarae* in constant conditions. HW and LW and/or their falling arrows indicate the expected times of high or low water. SS and SR indicate sunset and sunrise (Evans, 1976).

one of the places where it lives. When I worked with the animal years ago, it bore the more romantic scientific name of *Hadenoecus subterraneus*. We collected our specimens in Mammoth Cave in Kentucky, placed them in constant conditions, and found a fine nocturnal circadian rhythm (Reichle *et al.*, 1965).

The work has been repeated by Simon (1973), confirmed, and expanded. She carried out her study in an old mine that had been converted into a seismic observatory and was now sealed off with heavy steel doors and illuminated only for short intervals every 1–3 months when the strainmeters were serviced. The temperature remained steadily constant at 4.5°C, and the humidity at 90–95%; the surrounding darkness was the kind known only to spelunkers. The very sensitive strainmeters recorded the earthtides (described in Chapter 1) created by luni-solar attraction as the earth rotated; actographs next to them monitored cricket activity.

The earthtides repeated regularly with a semidiurnal inequality pattern, and, to everyone's surprise, in two out of three experiments, periodogram analysis indicated the same for the cave cricket! Unfortunately, the published figures are too small and over-inked to make reproduction worth while here. I have found no follow-up work in the literature.

Literature Cited

Aldrich, J. 1979. In situ recordings of oxygen consumption rhythm in *Carcinus maenas*. *Comp. Biochem. Physiol.*, 64A: 279–282.

Arudpragasam, K.D. and Naylor, E. 1964. Gill ventilation volumes, oxygen consumption and respiratory rhythms in *Carcinus maenas*. *J. Exp. Biol.*, 41: 309–321.

Atkinson, R.J. and Parsons, A.J. 1973. Seasonal patterns of migration and locomotor rhythmicity in populations of *Carcinus*. *Neth. J. Sea Res.*, 7: 81–93.

Barnwell, F.H. 1966. Daily and tidal patterns of activity in individual fiddler crabs (Genus *Uca*) from the Woods Hole region. *Biol. Bull.*, 130: 1–17.

Beentjes, M.P. and Williams, B.G. 1986. Endogenous circatidal rhythmicity in the New Zealand cockle *Chione stutchburyi*. *Mar. Behav. Physiol.*, 12: 171–189.

Beer, C.G. 1959. Notes on the behaviour of two estuarine crab species. *Trans. R. Soc. NZ*, 86: 197–203.

Bennett, M.F. 1954. The rhythmic activity of the quahog, *Venus mercenaria* and its modifications by light. *Biol. Bull.*, 107: 174–191.

Bennett, M.F., Shriner, J. and Brown, R.A. 1957. Persistent tidal cycles of spontaneous motor activity in the fiddler crab, *Uca pugnax*. *Biol. Bull.*, 112: 267–275.

Brown, F.A. 1954. Persistent activity rhythms in the oyster. *Am. J. Physiol.*, 178: 510–514.

Brown, F.A., Bennett, M.F. and Webb, H.M. 1954. Daily and tidal rhythms of O_2-consumption in fiddler crabs. *J. Cell. Comp. Physiol.*, 44: 477–506.

Brown, F.A., Bennett, M.F., Webb, H.M. and Ralph, C.L. 1956a. Persistent daily, monthly and 27-day cycles of activity in the oyster and quahog. *J. Exp. Zool.*, 131: 235–262.

Brown, F.A., Brown, R.A., Webb, H.M., Bennett, M. and Shriner, J. 1956b. A persistent tidal rhythm of locomotor activity in *Uca pugnax*. *Anat. Rec.*, 125: 613–614.

Davids, E. 1964. The influence of suspensions of micro-organisms of different concentrations on the pumping and retention of food by the mussel (*Mytilus edulis*). *Neth. J. Sea Res.*, 2: 233–249.

Dowse, H.B. and Palmer, J.D. 1990. Evidence for ultradian rhythmicity in an intertidal crab. In: Hayes, D.K., Pauly, J. and Reiter, R. (Eds), *Chronobiology: Its Role in Clinical Medicine, General Biology, and Agriculture*, pp. 691–697. Liss-Wiley, New York.

Dowse, H.B. and Palmer, J.D. 1992. Comparative studies of tidal rhythms. XI. Ultradian and circalunidian rhythmicity in four species of semiterrestrail, intertidal crabs. *Mar. Behav. Physiol.*, 21: 105–119.

Dowse, H.B. and Ringo, J.M. 1987. Further evidence that the circadian clock is a population of coupled ultradian oscillators. *J. Biol. Rhythms*, 2: 65–76.

Enright, J.T. 1965. The search for rhythmicity in biological time-series. *J. Theoret. Biol.*, 8: 426–468.

Evans, J.W. 1975. Growth and micromorphology of two bivalves exhibiting non-daily growth lines. Pp. 119–134. In: Rosenberg, G.D. and Runcorn, S.K. (Eds). *Growth Rhythms and the History of the Earth's Rotation*, pp. 119–134. John Wiley & Sons, New York.

Evans, W.G. 1976. Circadian and circatidal locomotory rhythms in the intertidal beetle *Thalassotrechus barbarae*: Carabidae. *J. Exp. Mar. Biol. Ecol.*, 22: 79–90.

Fielder, D.R. and Jones, M.B. 1978. Observations of feeding behaviour in two New Zealand mud crabs (*Helice crassa* and *Macrophthalmus hirtipes*). *Mauri Ora*, 6: 41–46.

Fingerman, M., Nagabhushanam, R. and Philpott, L. 1961. Physiology of the melanophores in the crab *Sesarma reticulatum*. *Biol. Bull.*, 120: 337–347.

Fish, J.D. and Fish, S. 1972. The swimming rhythm of *Eurydice pulchra*, and a possible explanation of intertidal migration. *J. Exp. Mar. Biol. Ecol.*, 8: 195–200.

Gibson, R.N. 1965. Rhythmic activity in littoral fish. *Nature*, 207: 544–545.

Gibson, R.N. 1976. Comparative studies on the rhythms of juvenile flatfish. In: DeCoursey, P.J. (Ed.), *Biological Rhythms in the Marine Environment*, pp. 199–213. Univ. South Carolina Press, South Carolina.

Gibson, R.N. 1982. Recent studies on the biology of intertidal fishes. *Ocean. Mar. Biol. Ann. Rev.*, 20: 363–414.

Gibson, R.N. 1992. Tidally synchronized behaviour in marine fishes. In: M. A. Ali (Ed.), *Rhythms in Fishes*, pp. 63–82. Plenum Press, London.

Goodwin, B. 1963. *Temporal Organization in Cells*. Academic Press, San Diego.

Gwinner, E. 1974. Testosterone induces "splitting" of circadian locomotor activity in birds. *Science*, 185: 72–74.

Happey-Wood, C.M. and Jones, P. 1988. Rhythms of vertical migration and motility in intertidal benthic diatoms with particular reference to *Pleurosigma angulatum*. *Diatom Res.*, 3: 83–93.

Harris, G.J. and Morgan, E. 1984. Rhythms of locomotion and oxygen consumption in the estuarine amphipod *Corophium volutator*. *Chronobiol. Int.*, 1: 21–25.

Harris, G.J. and Morgan, E. 1986. Seasonal and semi-lunar modulation of the endogenous swimming rhythm in the estuarine amphipod *Corophium volutator*. *Mar. Behav. Physiol.*, 12: 303–314.

Hastings, M.H. 1981a. The entraining effect of turbulence on the circatidal activity rhythm and its semi-lunar modulation in *Eurydice pulchra*. *J. Mar. Biol. Ass. UK*, 61: 151–160.

Hastings, M.H. 1981b. Semilunar variations of endogenous circatidal rhythms of activity and respiration in the isopod *Eurydice pulchra*. *Mar. Ecol. Prog. Ser.*, 4: 85–90.

Herdman, E.C. 1924. Notes on dinoflagellates and other organisms causing discoloration of the sand at Port Erin. Part III. *Proc. Trans. Lpool. Biol. Soc.*, 38: 58–63.

Hoffman, K. 1971. Splitting of the circadian rhythm as a function of light intensity. In: Menaker, M. (Ed.), *Biochronometry*, pp. 134–146. Natl. Acad. Sci., Washington, D.C.

Holmström, W.E. and Morgan, E. 1983. Variation in the naturally occurring rhythm of the estuarine amphipod, *Corophium volutator*. *J. mar. biol. Ass. UK*, 63: 833–850.

Holst, E.V. 1939. Die relative Koordination als Phänomen und als Methode zentralnervöser Funktiibsabaktse, *Ergebn. Physiol.*, 42: 228–306.

Hutchinson, D.N. 1988. Behavioural and physiological circatidal rhythms in the New Zealand cockle, *Chione stutchburyi*. MSc. Thesis, University of Otago, New Zealand.

Jones, D.A. and Naylor, E. 1970. The swimming rhythm of the sand beach isopod *Eurydice pulchra*. *J. Exp. Mar. Biol. Ecol.*, 4: 188– 199.

Jorgensen, C.B. 1960. Efficiency of particle retention and rate of water transport in undisturbed lamellibranchs. *J. Cons. Perma. Int. Explor. Mer.*, 26: 94–116.

Lutz, R.A. and Rhoads, D.C. 1977. Anaerobiosis and a theory of growth line formation. *Science*, 198: 1222–1227.

Morgan, E. 1965. The activity rhythm of the amphipod *Corophium volutator* and its possible relationship to changes in hydrostatic pressure associated with the tides. *J. Anim. Ecol.*, 34: 731–746.

Morton, B.S. 1971. The diurnal rhythm and tidal rhythm of feeding and digestion in *Ostrea edulis*. *Biol. J. Linn. Soc.*, 3: 329–342.

Naylor, E. 1958. Tidal and diurnal rhythms of locomotory activity in *Carcinus maenas*. *J. Exp. Biol.*, 35: 602–610.

Naylor, E. 1960. Locomotory rhythms in *Carcinus maenas* from non-tidal conditions. *J. Exp. Biol.*, 37: 481–488.

Naylor, E. 1961. Spontaneous locomotor rhythm in mediterranean *Carcinus*. *Pubbl. Staz. Zool. Napoli*, 32: 58–63.

Naylor, E. 1962. Seasonal changes in a population of *Carcinus maenas* in the littoral zone. *J. Anim. Ecol.*, 31: 601–609.

Naylor, E. 1963. Temperature relationships of the locomotor rhythm of *Carcinus*. *J. Exp. Biol.*, 40: 669–679.

Northcott, S.J. 1991. A comparison of circatidal rhythmicity and entrainment by hydrostatic pressure cycles in the rock goby, *Gobius paganellus* and the shanny, *Liphophrys pholis*. *J. Fish. Biol.*, 39: 25–33.

Northcott, S.J., Gibson, R.N. and Morgan, E. 1990. The persistence and modulation of endogenous circatidal rhythmicity in *Lipophrys pholis*. *J. Mar. Biol. Ass. UK*, 70: 815–827.

Page, T.L. 1989. Masking in invertebrates. *Chronobiol. Int.*, 6: 3–11.

Palmer, J.D. 1963. "Circa-tidal" activity rhythms in fiddler crabs. Effect of light intensity. *Biol. Bull.*, 125: 387.

Palmer, J.D. 1964a. A persistent, light-preference rhythm in the fiddler crab, *Uca pugnax* and its possible adaptive significance. *Amer. Nat.*, 98: 431–434.

Palmer, J.D. 1964b. Comparative studies of avian persistent rhythms: spontaneous change in period length. *Comp. Biochem. Physiol.*, 21: 273–282.

Palmer, J.D. 1967. Daily and tidal components in the persistent rhythmic activity of the crab, *Sesarma*. *Nature*, 215: 64–66.

Palmer, J.D. 1973. Tidal rhythms: the clock control of the rhythmic physiology of marine organisms. *Biol. Rev.*, 48: 377–418.

Palmer, J.D. 1974. *Biological Clocks in Marine Organisms: The Control of Physiological and Behavioral Tidal Rhythms*. John Wiley & Sons, New York.

Palmer, J.D. 1975. Biological clocks of the tidal zone. *Sci. Amer.*, 232: 70–79.

Palmer, J.D. 1976. Clock-controlled vertical migration rhythms in intertidal organisms. In: P.J. DeCoursey (Ed.), *Biological Rhythms in the Marine*

Environment, pp. 239–255. University of South Carolina Press, South Carolina.

Palmer, J.D. 1988. Comparative studies of tidal rhythms. VI. Several clocks govern the activity of two species of fiddler crabs. *Mar. Behav. Physiol.*, 13: 201–219.

Palmer, J.D. 1989a. Comparative studies of tidal rhythms. VII. The circalunidian locomotor rhythm of the brackish-water fiddler crab, *Uca minax*. *Mar. Behav. Physiol.*, 14: 129–143.

Palmer, J.D. 1989b. Comparative studies of tidal rhythms. VIII. A translocation experiment involving circalunidian rhythms. *Mar. Behav. Physiol.*, 14: 231–243.

Palmer, J.D. 1990a. Comparative studies of tidal rhythms. X. A dissection of the persistent activity rhythms of the crab, *Sesarma*. *Mar. Behav. Physiol.*, 17: 177–187.

Palmer, J.D. 1990b. The rhythmic lives of crabs. *BioScience*, 40: 352–358.

Palmer, J.D. and Round, F.E. 1965. Persistent, vertical-migration rhythms in benthic microflora. I. The effect of light and temperature on the rhythmic behaviour of *Euglena obtusa*. *J. Mar. Biol. Ass. UK*, 45: 567–582.

Palmer, J.D. and Round, F.E. 1967. Persistent, vertical-migration rhythms in benthic microflora. VI. The tidal and diurnal nature of the rhythm in the diatom *Hantzschia virgata*. *Biol. Bull.*, 132: 44–55.

Palmer, J.D., Udry, J.R. and Morris, N.M. 1982. Diurnal and weekly, but no lunar rhythms in human copulation. *Human Biol.*, 54: 111–121.

Palmer, J.D. and Williams, B.G. 1986a. Comparative studies of tidal rhythms. I. The characterization of the activity rhythm of the pliant-pendulum crab, *Helice crassa*. *Mar. Behav. Physiol.*, 12: 197–207.

Palmer, J.D. and Williams, B.G. 1986b. Comparative studies of tidal rhythms. II. The dual clock control of the locomotor rhythms of two decapod crustaceans. *Mar. Behav. Physiol.*, 12: 269–278.

Palmer, J.D. and Williams, B.G. 1987. Comparative studies of tidal rhythms. III. Spontaneous splitting of the peaks of crab locomotory rhythms. *Mar. Behav. Physiol.*, 13: 63–75.

Palmer, J.D. and Williams, B.G. 1993a. Comparative studies of tidal rhythms. XII. Persistent photoaccumulation and locomotor rhythms in the crab, *Cyclograpsus lavauxi*. *Mar. Behav. Physiol.*, 22: 119–129.

Palmer, J.D. and Williams, B.G. 1993b. An organismic tidal rhythm with a peculiar phenotype. In: Adrich, J. (Ed.), *Quantified Phenotypic Responses in Morphology and Physiology*, pp. 121–127. JAPAGA, Ashford, Ireland.

Palmer, R.E. 1980. Behavioural and rhythmic aspects of filtration in the bay scallop *Argopecten irradians concentricus* and the oyster *Crassostre virginica*. *J. Exp. Mar. Biol. Ecol.*, 45: 273–295.

Pavlidis, T. 1971. Populations of biochemical oscillators as circadian clocks. *J. Theor. Biol.*, 33: 319–338.

Pittendrigh, C.S. and Daan, S. 1976a. A functional analysis of circadian pacemakers in nocturnal rodents. I. The stability and lability of spontaneous frequency. *J. Comp. Physiol.*, 106: 223–252.

Pittendrigh, C.S. and Daan, S. 1976b. A functional analysis of circadian pacemakers in nocturnal rodents. V. Pacemaker structure: a clock for all seasons. *J. Comp. Physiol.*, 106: 333–355.

Rao, K.P. 1954. Tidal rhythmicity of rate of water propulsion by *Mytilus*, and its modification by transplantation. *Biol. Bull.*, 106: 353–359.

Reichle, D.E., Palmer, J.D. and Park, O. 1965. Persistent, rhythmic locomotory activity in the cave cricket, *Hadenoecus subterraneus*, and its ecological significance. *Am. Midl. Nat.*, 74: 57–66.

Reid, D.G. and Naylor, E. 1985. Free-running, endogenous semilunar rhythmicity in a marine isopod crustacean. *J. Mar. Biol. Ass. UK*, 65: 85–91.

Resko, J.A. and Eik-Nes, K.B. 1966. Diurnal testosterone levels in peripheral plasma of human male subjects. *J. Clin. Endocrin.*, 26: 573–576.

Richardson, C.A., Crisp, D.J. and Runham, N.W. 1980. An endogenous rhythm in shell deposition in *Cerastoderma edule*. *J. Mar. Biol. Ass. UK*, 60: 991–1004.

Round, F.E. and Palmer, J.D. 1966. Persistent, vertical-migration rhythms in benthic microflora. II. Field and laboratory studies of diatoms from the banks of the River Avon. *J. Mar. Biol. Ass. UK*, 46: 191–214.

Seiple, W. 1981. The ecological significance of the locomotor activity rhythms of *Sesarma cinereum* and *Sesarma reticulatum*. *Crustaceana*, 40: 5–15.

Simon, R.B. 1973. Cave cricket activity rhythms and the earth tides. *J. Interdiscipl. Cycle Res.*, 4: 31–39.

Southern, A.L., Gordon, F.G., Tochimota, S., Pinzon, G., Lane, D.R. and Stypulkowski, W. 1967. Mean plasma concentration, metabolic clearance and basal plasma production rates of testosterone in normal young men and women using a constant infusion procedure: effect of time of day and plasma concentration on the metabolic clearance rate of testosterone. *J. Clin. Endocrin.*, 27: 686–694.

Taylor, W.R. 1964. Light and photosynthesis in intertidal benthic diatoms. *Helgol. Wiss. Merresunters*, 10: 29–37.

Taylor, W.R. and Palmer, J.D. 1963. The relationship between light and photosynthesis in intertidal benthic diatoms. *Biol. Bull.*, 125: 395.

Thompson, R.J. and Bayne, B.L. 1972. Active metabolism associated with feeding in the mussel *Mytilus edulis*. *J. Exp. Mar. Biol. Ecol.*, 9: 111–124.

Turek, D.W., Earnest, D.J. and Swann, J. 1982. Splitting of the circadian rhythm of activity in hamsters. In: Aschoff, J., Dann, S. and Groos, G. (Eds), *Vertebrate Circadian Systems*, pp. 203–214. Springer-Verlag, Berlin.

Underwood, H. 1977. Circadian organization in lizards: the role of the pineal organ. *Science*, 195: 587–589.

Warman, C.G., O'Hare, T.J. and Naylor, E. 1991. Vertical swimming in wave-induced currents as a control mechanism of intertidal migration by a sand-beach isopod. *Mar. Biol.*, 111: 49–50.

Warman, C.G., Reid, D.G. and Naylor, E. 1993. Variation in the tidal migratory behaviour and rhythmic light-responsiveness in the shore crab, *Carcinus maenas*. *J. Mar. Biol. Ass. UK*, 73: 355–364.

Webb, H.M. 1971. Effects of artificial 24-hour cycles on the tidal rhythm of activity in the fiddler crab, *Uca pugnax*. *J. Interdiscipl. Cycle Res.*, 2: 191–198.

Wells, J.W. 1963. Coral growth and geochronometry. *Nature*, 197: 948–950.

Williams, B.G. 1991. Comparative studies of tidal rhythms. V. Individual variation in the rhythmic behaviour of *Carcinus maenas*. *Mar. Biol. Physiol.*, 19: 97–112.

Williams, B.G. 1995. Tidal biological clocks and their diel counterparts. In:

Hartnoll, R.G. and Hawkins, S.J. (Eds), *Marine Biology—A Port Erin Perspective*, in press. Immel Publishing Co., London.

Williams, B.G. and Naylor, E. 1967. Spontaneously induced rhythm of tidal periodicity in laboratory-reared *Carcinus. J. Exp. Biol.*, 47: 229–234.

Williams, B.G., Naylor, E. and Chatterton, T.D. 1985. The activity patterns of New Zealand mud crabs under field and laboratory conditions. *J. Exp. Mar. Biol. Ecol.*, 89: 269–282.

Williams, B.G. and Palmer, J.D. 1988. Comparative studies of tidal rhythms. IV. Spontaneous frequency changes and persistence in the locomotor rhythm of an intertidal crab. *Mar. Behav. Physiol.*, 13: 315–332.

Williams, B.G., Palmer, J.D. and Hutchinson, D.N. 1993. Comparative studies of tidal rhythms. XIII. Is a clam clock similar to those of other intertidal animals? *Mar. Behav. Physiol.*, 24: 1–14.

Williams, J.A. 1985. An endogenous tidal rhythm of blood-sugar concentrations in the shore crab *Carcinus maenas. Comp. Biochem. Physiol.*, 81A: 627–631.

4

Phase Setting
and Entrainment

The phase of the tides can vary greatly over relatively short distances of coastline. The rhythmic behavioral patterns of the animals examined from each different locale reflect the phase of the tides at their collection site. Their rhythms are said to be *entrained* (i.e., synchronized) by the tides. One need not be a rocket scientist to suspect that there must be some physico-chemical parameters of the tides that actually set this phase. Finding them is the subject of this chapter.

The discussion could be organized by phase-setting factor (as I have done in a previous book: Palmer, 1974) or by organism. Feedback from previous monograph readers suggested the latter might be better because marine chronobiological laboratories tend to concentrate on favorite study subjects. Thus, this is what I will use here.

The Green Shore Crab (Carcinus)

The pioneering work was begun by Williams and Naylor (1969) who examined the most obvious choice first: inundation cycles. The first subjects they chose were the so-called "dock" crabs described in Chapter 3 — animals that lived on the atidal underside of a floating raft and thus displayed only circadian rhythms in constant conditions. They were caged and half of them placed at midtide level on the shore, and the other half sunk subtidally where they remained always covered by the ocean, but were, of course, exposed to the physical and chemical alterations created by the periodic changes in the current, and in the height of the water column overhead. After a few days, the animals' locomotory output was tested in the laboratory where both groups were found to show persistent tidal rhythms. Peak activity was phased to the times of expected high tides. Thus, the axiomatic appeared to be proved, periodic exposure to air would initiate and entrain a tidal rhythm, but was not the only condition that would do so as seen by the response of the caged crabs positioned subtidally.

After one works with organismic rhythms for a short while one learns not to be smug, even over experimental results that most everyone would expect to get. Wisely, Williams and Naylor designed another experiment to test the role of periodic exposure to air. Fresh, rhythmic crabs were subject for 5 days to 12.4-h cycles consisting of 6.2 h in air and 6.2 h under a few centimeters of water. The especially wise part of the design was to hold the temperature at 19°C in both conditions (thus eliminating temperature-change as a variable). As the crabs were already rhythmic, the immersion cycles were offered out of phase with the animals' activity rhythm. During the immersion phase the animals were active, and during exposure to air they remained still. To someone looking at the data logger, it appeared that the rhythm had been instantaneously inverted (Fig. 4-1, top). On the sixth day the crabs were switched to constant moist air held at 19°C: under this condition it was found that the existing rhythm had been cancelled out and the crabs had become arrhythmic! The most obvious entraining cycle, periodic inundation had not entrained.

Williams and Naylor (1969) delved deeper. Using dock crabs this time, they repeated the inundation cycles, but held the water temperature at 13°C, and the air temperature at 17°C or 24°C. Activity peaks appeared during the cool intervals. After ten repeats of these cycles, which provided a 4°C or 11°C temperature differential, the crabs were switched to constant conditions where it was found that the rhythm persisted, with peak activity phased to the times of the 13°C inundation.

The last rendition of the same design theme simply eliminated the water: the 13°C temperature phase was provided as cool air. During 5 days of treatment, the dock crabs mainly confined their activity bursts to the colder temperatures, and this entrained phase persisted in constant conditions (Fig. 4-1, bottom).

This nice series of experiments demonstrates that the entraining role of the water in this case was just to deliver cool temperature pulses.

As described above, caged dock crabs held subtidally on the ocean bottom for a few days had a tidal rhythm instilled in them. Tide-created hydrostatic-pressure changes were the suspected cause, so an apparatus was created to deliver square-wave pressure cycles in the laboratory. Arrhythmic crabs were subjected to varying numbers of cycles of 6 h at atmospheric

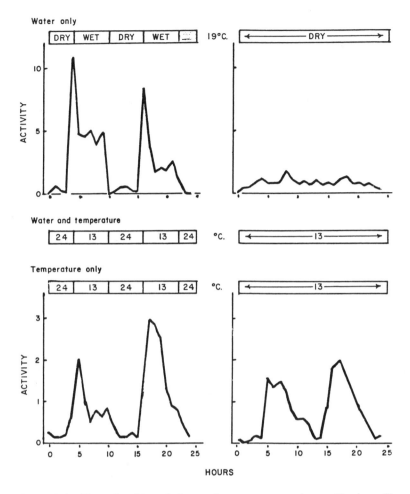

Figure 4-1 The role of inundation and temperature cycles on *Carcinus*. *Top left:* Rhythmic crabs were exposed to 12.4-h inundation cycles offered 180° out of phase from their own. The temperature was held constant at 19°C. The crabs appeared to have been entrained to the times of inundation. *Top right:* When released into constant conditions the crabs were found to be arrhythmic. *Middle:* The experiment was repeated, but this time the air and water temperature differed as indicated. This procedure successfully re-entrained the rhythm. *Bottom:* In this variation the temperature cycle was offered alone. It entrained the rhythm to its phase (drawn from the data of Williams and Naylor, 1969).

pressure, alternating with 6 h at some increased pressure. During exposure to the pressure cycles activity bursts built up under the increased pressure phase, and this pattern was conserved as a tidal rhythm in constant conditions thereafter. The results of one such experiment are seen in Fig. 4-2; note that for all practical purposes, a semidiurnal inequality of

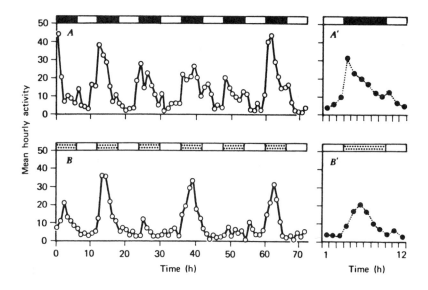

Figure 4-2 The induction of a tidal rhythm in *C. maenas* by subjecting six arrhythmic crabs to a hydrostatic pressure cycle consisting of 6 hours at 1.0 atmosphere, alternating with 6 hours at 1.6 atmospheres for 6 cycles (A) and then transferring them to a constant pressure of 1.0 atmosphere (B). Note that the treatment produced a persistent rhythm with a semidiurnal inequality of peak amplitude. The blackened segments of the overhead bars represent the times of high pressure; the stippled sections represent the times of *expected* high pressure. Form estimates of the responses are seen to the right of the figure (Naylor & Atkinson, 1972).

amplitude appears to have been molded into the persistent rhythm! Hydrostatic pressure as low as only 1.1 atmospheres, offered for six cycles, was found adequate to produce these results. The greater the number of times the pressure cycle was repeated, the greater the amplitude of the resulting rhythm (Naylor *et al.*, 1971; Naylor & Atkinson, 1972). An apparatus was then built that would produce sinusoidal pressure cycles (Reid *et al.*, 1989). Single, arrhythmic crabs were exposed for 4 days to 12.4-h cycles of alternating pressures between 1.0 and 1.5 atmospheres (equal to 5 meters of water overhead), and then studied in constant conditions for 4 days. The treatment initiated rhythmicity and set its phase to the times of high pressure. There is no mention of an inequality of alternating peaks (Reid & Naylor, 1990). The tide-associated rhythm in blood-sugar concentration can also be re-started and its phase set by laboratory-imposed pressure cycles: the low blood sugar phase locks with low pressure (Williams, 1985).

Another extreme change brought to intertidal dwellers by each flood tide is a hyper-osmotic dunking. The first investigation of the role of salinity change on a tidal rhythm was carried out by Taylor and Naylor

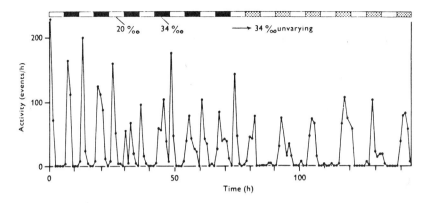

Figure 4-3 The locomotor activity in constant dim red light and 13°C of a single, previously arrhythmic, *Carcinus* during 78 hours' exposure to salinity-change cycles (20–34‰, indicated by the open and solid segment of the overhead bar, respectively), followed by 66 h in full seawater. The stippled segments of the post-treatment supertending bar signify the expected times of high salinity (Taylor & Naylor, 1977).

(1977). They subjected arrhythmic animals to square-wave cycles of salinity change consisting of 6-h stints in 20‰ seawater, alternating with 6-h stays in full seawater (34‰). Such a change might be expected in an estuarine habitat. Fig. 4-3 shows that both steps up and steps down in the ambient salt content sparked bursts of crab activity, giving the overt impression that the animals had been entrained to a 6-h locomotory cycle. This was not the case however; when the crabs were then released into constant conditions that included inundation in full seawater, the only activity peaks to survive were those that had been associated with the salty tops of imposed square waves during pretreatment. The intervals of diluted sea water had apparently just temporarily provoked locomotion, and played no role in setting the tidal clock. This supposition was proved in a later experiment. As will be described in the last chapter, the activity rhythms of *Carcinus* are influenced by, for what will now be called, neurosecretory clocks in the eyestalks. When the stalks are cut off, the locomotor rhythm is lost. As would be expected, imposed salinity cycles do not resuscitate lost locomotor rhythms in eyestalkless crabs, but activity is caused to increase when each low-salinity pulse is offered. This augmentation could not be a result of dilute seawater acting directly on the eyestalk "clocks" because those timepieces are in the wastebin (Bolt & Naylor, 1985).

Part of this experiment was repeated, this time in an improved version where the crabs were exposed to more natural 12.4-h, *sinusoidal*, salinity-change cycles. In this inquiry the subjects were collected in the winter from a subtidal locale to which they migrate late each autumn. Because the animals in this habitat are exposed to pressure cycles they would be expected to be rhythmic, but when brought into the lab it was found that

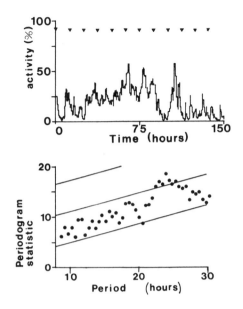

Figure 4-4 *Top:* The group response of eight winter-collected *Carcinus* maintained in 34‰ seawater, 15°C, and constant darkness for 6 days. The solid triangles indicate the times of high tide in the crabs' natural habitat. *Bottom:* Periodogram of the same data indicating an absence of significant periodicity (Bolt & Naylor, 1985).

they were not (the low temperature may be the culprit here: Atkinson and Parsons (1973) found that *Carcinus* is not rhythmic below 8°C). Anyway, the crabs were first observed in constant conditions for 6 days to assure their arrhythmicity: while their activity was often organized into spikes, a rhythm could not be demonstrated by periodogram analysis (Fig. 4-4). (It is curious that the circadian component was not initiated by the step up to 15°C from winter seawater temperature used in the laboratory. One-time temperature steps are known to initiate circadian rhythmicity in other organisms. Also, Williams and Naylor (1967) have demonstrated that a single step up in temperature from 4°C to 15°C will initiate a tidal rhythm in *Carcinus*. However, in this case these stimuli appeared ineffective.) So winter crabs were then subjected to sinusoidal salinity-change cycles, a treatment that caused them to display (Fig. 4-5) an overt activity cycle with peaks being expressed in synchrony with the troughs of imposed low salinity (Bolt & Naylor, 1985) — rather than at both increased and decreased salinity changes. Note that in conditions like these, where salinity change was offered *gradually*, high salinity produced only a couple of spikes of activity in the beginning of the treatment. Considering the result depicted in Fig. 4-3, where exposure to low salt was wholly ineffective as an entrainer, it would seem obvious that the *apparent* rhythm shown in

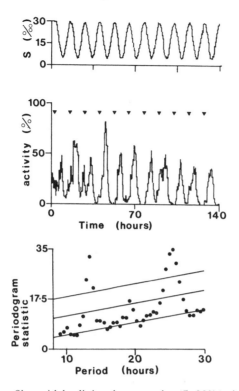

Figure 4-5 *Top:* Sinusoidal salinity-change cycles (5–30‰) time so that the peaks of highest salinity corresponded to the expected times of high tide. *Middle:* The collective activity of eight *Carcinus*, freshly collected in the winter, and exposed to constant darkness, 15°C, and the salinity-change cycle (solid triangles signify times of high salinity and what would be high tides). *Bottom:* Periodogram analysis showing the presence of significant peaks at 13.5 h and 25.6 h (looking at the raw data, my guess would be that the latter is just a supermultiple of the smaller value) (Bolt & Naylor, 1985).

Fig. 4-5 should have been tested for persistence in constant conditions; however, it was not.

This omission was corrected in a subsequent experiment, but the design was changed somewhat. In these observations, crabs were collected intertidally in the summer and thus possessed fully developed tidal rhythms. The salinity-change cycle was therefore positioned so that the times of high salinity were offered antiphase to the times of expected high tides. Thus the low salinity offerings (which are known to elicit activity, but not to initiate rhythms) were presented in synchrony with the times of expected high tides (times when the clock dictates activity), meaning that the crabs, during the attempted-entraining cycles, should be doubly stimulated to run at those times. For the first seven cycles this was the case (Fig. 4-6), but then low amplitude activity peaks began to form at the times

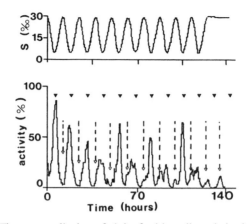

Figure 4-6 The group display of eight freshly collected rhythmic *Carcinus* exposed for 135 h to 12.4-h, 5–30‰ salinity-change cycles, at 15°C, and in constant darkness; and then switched to a constant salinity of 30‰ for 16 h. The dilute seawater phase of the salinity cycle was presented in synchrony with the times of expected high tide (solid triangles). The falling arrows signify the peaks of high salinity (Bolt & Naylor, 1985).

of high salinity, and it was these that persevered thereafter in constant conditions (Bolt & Naylor, 1985).

In the above work, only the group response of eight crabs was used. The identical experiment has since been repeated using single, rhythmic crabs. Salinity cycles (10‰ alternating with 35‰, offered antiphase to the times of expected high tides (i.e., low salinity at high tide), were offered for 4 days and the crabs then moved to constant conditions. No mention is made of what happened to 56 of the 72 crabs used, but 16 of them displayed *c*.6.2-h persistent cycles, with peaks approximately phased to the expected times of high tide *and* high salinity (Reid & Naylor, 1990), rather than just a rephasing as found by Bolt and Naylor (1985) and described in the paragraph above.

You will remember from Chapter 3 that *Carcinus mediterraneus* from the essentially atidal Mediterranean Ocean show only circadian rhythms in the laboratory. Out of curiosity, this species was subjected to 12.4-h high/low salinity cycles to see if a tidal rhythm could be imposed on it. It could not (Warman *et al.*, 1991).

As it turns out, salinity-change *cycles* are not necessary to initiate tidal rhythmicity in *Carcinus*. Arrhythmic, winter-collected crabs maintained in otherwise constant conditions were simply immersed in seawater diluted to 10% of normal (thus the observation depicted in Fig. 4-4 can serve as a control for this experiment), and this initiated ". . . weak circatidal and stronger circadian rhythms" (Fig. 4-7) (if that interpretation is correct, this is the first circadian rhythm ever started by a salinity change). The subtending periodogram in this figure suggests significant periods of 13.6

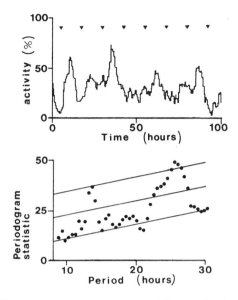

Figure 4-7 The group activity pattern of eight, previously arrhythmic, *Carcinus* after being exposed to constant darkness, 15°C, and 10‰ seawater. Solid triangles indicate the times of expected high tide. Periodogram analysis of the data indicates "the presence of weak circatidal [13.6 h] and circadian [25.4 h] rhythmicity" (Bolt & Naylor, 1985).

and 25.4 h (Bolt & Naylor, 1985). Curious, is it not, that in the work previous to this low salinity would only stimulate activity, but would not initiate a persistent rhythm or set its phase!

The response to salinity change is even more complicated: the point in the light/dark cycle at which the crabs were transferred to low salinity was found to play a very important role in setting the phase and determining the form of the subsequent rhythmic display. Here again, arrhythmic, winter-collected crabs were used, but first they were exposed (while immersed in full seawater) for two days to natural day/night cycles. These cycles would be expected to start, and set the phase, of the solar-day component. I should mention that in this series, individual animals were used and "as . . . usual 25% showed no activity and another 25% displayed only random activity. . . ." After this pretreatment, the crabs were switched to constant darkness and a step down to 20% seawater. Separate groups of animals were transferred at 0600 h, 1200 h, 1800 h, and 2400 h. Approximately 12.4 h after the 0600 h, 1200 h, and 1800 h transfers were made, an activity peak materialized. In the 0600-h and 1800-h transfers, this first peak was followed at circatidal intervals by subsequent peaks. But in animals transferred at 1200 h, this first peak was followed by others at circadian "or twice tidal" intervals, rather than at circatidal intervals. In the case of the transfer made at 2400 h, the first peak, and subsequent ones,

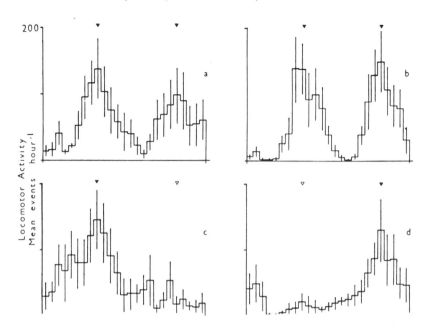

Figure 4-8   Twenty-five-hour form-estimate curves for groups of up to 12 *Carcinus* (previously arrhythmic but subjected to a pretreatment of day/night cycles while immersed in full seawater at 10°C, for two days) exposed to constant darkness and 20% seawater at: (a) 1800 h, (b) 0600 h, (c) 1200 h, and (d) 2400 h, and studied for the next 78 hours. The open triangles represent the end of 12.4-h intervals starting after the transfer to 20% seawater; the solid ones indicate the times that coincided with expected night-time, and the open ones occurring during expected daytime. Skewers are ± one standard error (Reid & Naylor, 1989).

appeared at circadian or twice tidal intervals (Fig. 4-8). The authors of this work (Reid & Naylor, 1989) point out that after transfers at 1200 h and 2400 h, the missing tidal peaks were those that normally would have occurred during the daytime hours, and they speculate that the best explanation must be that the circadian component totally suppressed these peaks.

Further work along these lines produced the following. Rhythmic crabs were collected and used immediately for experimentation. At a time approximating expected low tide, 200 *Carcinus* were placed into individual actographs in constant darkness and immersed in 20% seawater. About 6 h later, at the time of the next expected high tide, as anticipated, activity peaks were displayed in those animals that were rhythmic. Then, seen in a few crabs, approximately 6 h after that peak a second one burgeoned, approximately 12.5 h after submerging the crabs in 20% seawater. Thus *two* peaks were exhibited in the first 12.5 h in constant conditions (Fig. 4-9), and this pattern was repeated, twice a day, for the next 4.5 days. Only 15 of

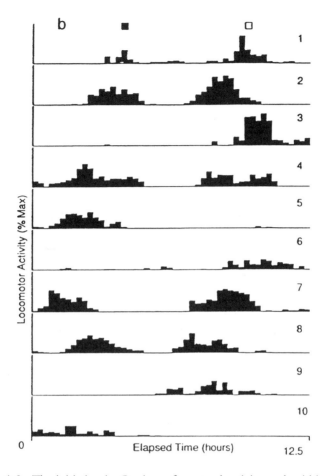

Figure 4-9 The initiation in *Carcinus* of a second activity peak within a single tidal interval. Slightly before time 0 on the ordinate on the first day of the experiment, this individual was submerged in 20% seawater and exposed to constant darkness. At the next time of expected high tide (indicated by the solid box overhead) an activity burst appeared. Then, 6 h later a second peak of activity sprung up (under the open box), producing the curious display of *two* peaks in one 12.5-h tidal interval. This display was repeated 10 more times. While the variability of the pattern is great, it seems possible to opine that the periods of the two peaks are different (Reid & Naylor, 1993).

the 200 crabs produced identifiable examples of this doubling of peaks. Although these patterns are ragged, there is a suggestion that the two peaks sometimes scan the 12.5-h interval at different rates. Reid and Naylor (1993), the authors of this work, report period differences that range between 2 and 26 min (mean = 13 min), but because of the short length of the data strings, and the imprecision of the patterns, only in four animals did the difference suggest statistical significance.

To confirm that it was "hypo-osmotic shock" that was responsible for the creation of the second peak, the same experimental design was used again with a small modification: 24 crabs were put into 20% seawater and constant darkness 3 h — rather than 6 h — after expected high tide. Thirteen of the crabs were either arrhythmic or showed only the typical rhythm phased to expected high tides, but the other 11 flaunted the second peak *c*.12.5 h after being submerged in dilute seawater. Because the dunking began 3 h earlier than in the previous design, the second peak arose only ≈3 hours — rather than 6 hours — after the peak at expected high tide. Thus, the authors conclude that dilute saltwater can, in some *Carcinus*, initiate the formation of peak at an unexpected time. (Just for the record, it is probably worth mentioning that not one, but two, variables were administrated simultaneously: 20% seawater and the onset of constant darkness).

Reid and Naylor contemplated whether they were dealing with a splitting of a normal tidal peak in two, or whether the low-salt diet galvanized a second clock to action. They favored the latter.

This whole series of results using salinity cycles and pulses is very difficult to interpret: many of the findings are contradictory. Hopefully, the Naylor group sometime will attempt to sort this out and share their interpretation with the rest of us.

The combined action of temperature-change and salinity-change cycles was then tested (Bolt *et al.*, 1989). Crabs were offered both cycles in their two possible cardinal phase relationships. A complicating factor was the use of fresh, summer-collected, *rhythmic* crabs. In the first experiment, 12.4 h temperature-change (10–20°C) cycles, and 5–30‰ salinity-change cycles were presented with low temperature and high salinity in phase — the configuration found in the natural habitat — but offered antiphase to the times of expected high tides. During 130 h of this treatment, the crab's activity pattern was characterized by a staccato of peaks: one every 6.2 h, with maxima approximately synchronized to low temperature/high salinity/low tide, and high tide/low salinity/ high tide (Fig. 4-10). When the crabs were then switched to "10°C and 30‰" [*sic*, but see Fig. 4-10] salinity for 50 h, it was found that the only activity bursts that survived came at the times of expected low temperature/high salinity/low tide. Thus, the crabs' normal rhythm had been rephased 180°, meaning that the continuously produced peaks seen at the times of expected high tide during the first 130 h of treatment were mainly temporary bursts of activity induced by exposure to low salinity, a result expected from the data depicted in Fig. 4-3.

Now the experiment was repeated again; this time low temperature was offered in phase with low salinity (no mention is made of how the phase of these combined artificial cycles related to that of the times of the expected high tides). During 70 h of attempted entrainment, activity peaks were generated only at the times of low temperature/low salinity (Fig. 4-11). The crabs were then released into constant 15°C and 15‰ salinity

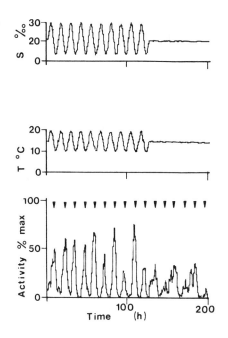

Figure 4-10 *Top:* Ten oscillations of a 5–30‰ salinity cycle followed by 50 h at 30‰. *Middle:* The oscillations of a 10–20°C temperature cycle followed by 50 h at 10°C. *Bottom:*The collective activity of eight *Carcinus* exposed to the combination of the above two cycles (and continuous dim red light) for 130 h, and in constancy thereafter. The solid triangles represent the actual, or, after the onset of constant conditions, the expected times of low temperature/high salinity (Bolt *et al.*, 1989).

for about 70 h. Here the major peaks tended to be expressed at the times of expected low temperature/low salinity; but the authors also found importance in the broadening left shoulders of these maxima and/or in the between-peak peaklets also expressed here. They wondered if these could have been produced by the previous periodic exposure to high salinity.

To satisfy their curiosity, additional experiments were conducted, but the design was somewhat changed: instead of using groups of crabs, individuals were used and they were first conditioned for two weeks in full seawater, 15°C, and a natural day/night cycle. This pretreatment would be expected to have insured the continuous running of any 24-h-based clocks. The crabs were then exposed for 4 days to the same artificial cycles used in the paragraph above with low temperature and low salinity presented in phase. As seen in Fig. 4-12, during the attempted entrainment interval, the main activity peaks formed at the times of low temperature/low salinity, while peaklets burgeoned around the times of high temperature/high salinity (the latter peaks had not been expressed in previous experiments [such as seen in Fig. 4-11]). But when the crabs were switched to constant

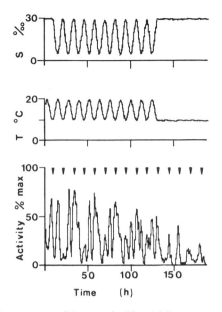

Figure 4-11   The same conditions as in Fig. 4.10 are presented here, except that here the solid triangles represent an altered phase relationship between the temperature and salinity cycles: low temperature is now in synchrony with low salinity. The solid triangles represent the actual, and then later the expected, times of low temperature/low salinity (Bolt *et al.*, 1989).

15‰ salinity and 15°C, the peaklets grew to be the dominant maxima. The results obtained in constant conditions with both of the animals displayed peaks with varying amplitudes, and a rhythm with a period of about 6.2 h. The authors concluded from these responses that temperature and salinity cycles entrain separate components of a crab's tidal clock. Such a conclusion is tempting, but loses some impact when the results of the final two experiments came in. In these, crabs were exposed for 130 h to *either* 6.2-h cycles of 5–30‰ salinity change; or 6.2-h, 10–20°C cycles of temperature change. In the former, activity bouts erupted in synchrony with low salinity, but the crabs were demonstrated to be arrhythmic when moved to constant conditions (Fig. 4-13). In the latter, activity peaks arose in phase with *every fourth* offering of low temperature, and their amplitudes damped rapidly with time. When switched to constant conditions, the activity was random to negligible (Fig. 4-13). Thus, interpretation of the results shown in Fig. 4-12, is very difficult, and the authors correctly conclude that further study would be worth while (Bolt *et al.*, 1989).

That additional work has been done. Arrhythmic crabs were *individually* subjected to 4 days of 12.4-h cycles of pressure and temperature, with high pressure being offered 6.2 h out of phase with low temperature. Of the 72 crabs so treated, 20 of those transferred to constant conditions

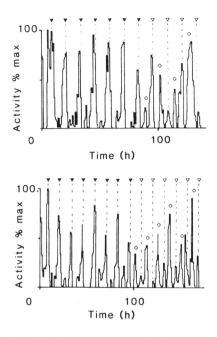

Figure 4-12 Again, artificial temperature and salinity cycles were offered, with low temperature in phase with low salinity (as in Figs 4.10 and 4.11), a relationship represented by the solid triangles. The two *Carcinus* shown have been stripped of their tidal rhythms by long-term storage in the laboratory. Clearly the crabs entrained to the times of low temperature/low salinity, but "peaklets" are also usually seen between these maxima. Then the crabs were moved to 15‰ and 15°C where the previous major peaks persisted at the times of expected low temperature/low salinity (open triangles), and the peaklets increased in amplitude and prominence and were in approximate phase with the expected times of high temperature/high salinity (indicated by the open diamonds) (Bolt *et al.*, 1989).

showed peaks of activity at the expected times of high pressure *and* low temperature (i.e., one peak was displayed approximately every 6.2 h); 17 crabs had phased only to the times of expected high pressure; 12 synchronized to the subjective times of low temperature, and 23 were inactive or their activity non-rhythmic. The same schedule of pretreatment was repeated using cycles of pressure and salinity: maximum pressure was offered antiphase with maximum salinity. When moved to constant conditions, 23 of the 80 subjects showed *c*.6.2-h cycles, with peaks coming at the expected times of both high pressure and high salinity. No mention is made of what the other 57 animals did (Reid & Naylor, 1990). Cycles of salinity and temperature were not tested, but Bolt *et al.* (1989) have tried that combination at least once and reported no incidence of *c*.6.2 h cycles.

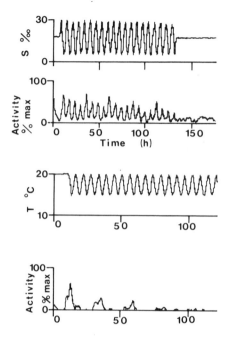

Figure 4-13 The group response of eight crabs (no mention is made whether they were arrhythmic or not) subjected to: *Top:* 6.2-h cycles of salinity change (5–30‰) at a constant temperature for 130 h, and then switched to 15‰ salinity for 50 h. *Bottom:* 6.2-h cycles of 10–20°C temperature change. In the upper design the crabs became arrhythmic in constant conditions; while, curiously, in the lower data set, in constant conditions peak activity arose at every fourth peak of high temperature for a while (Bolt *et al.*, 1989).

The underlying means of generating *persistent c.*6.2-h rhythms after pretreatment is unknown and difficult to explain. It seems logical that a half-tide period must be only a laboratory phenomenon, because if it occurred in nature — so that the crabs ran at both high and low tides ~ the animals would be twice as likely to fall victim to predators. It being a laboratory expression that can be created in arrhythmic crabs, it is curious that offering 6.2 h cycles of temperature alone, or salinity alone, did not produce the same effect, as was reported by Bolt *et al.* (1989), and shown here in Fig. 4-13. Perhaps if a larger number of subjects had been used, some individuals would have been found that would pick up the 6.2-h rhythms. The authors of this study ruminate that maybe more than one clock controls the tide-associated rhythms of this animal. I suppose that could be the case, for, as will be described in the last chapter, there is reason to believe that a clock exists in each eyestalk of *Carcinus*, and another clock may exist in the CNS. It may be that the eyestalks have been entrained by, say, temperature cycles and the CNS clock by, say, pressure cycles.

The possibility also exists that the treatments cause lunidian clocks to split their output.

The Estuarine Crab (Rhithropanopeus harrissii)

Ovigerous female crabs of this species, living in a part of an estuary not subjected to regular tides, release their larvae randomly throughout the day in constant conditions. However, if a sinusoidal salinity cycle (a 12.4-h alternation of 5‰ — 24.5‰ — 5‰) is added to otherwise constant conditions for 7–10 days, and the crabs then moved to atidal, 15‰ conditions, larvae are released from the sample group at the times of expected high salinity for the next 2.5 days. If instead, after the salinity cycle pretreatment, they are placed in 15‰ and a 13-h day/11-h night cycle, the crabs released only during the night-time expected high salinity peaks (Forward *et al.*, 1986).

The Portunid Crab (Liocarcinus holsatus)

This is a common crab of the British Isles. In the winter it lives subtidally, but between late spring and early autumn it enters the intertidal zone to feed. When collected from the tide zone during the summer, it exhibits a tide-associated activity rhythm in the laboratory in either constant illumination or darkness. Crabs collected from the subtidal zone do not.

Crabs stored in the laboratory for several weeks show no rhythmicity when tested using periodogram analysis. But after being exposed to six sinusoidal hydrostatic pressure cycles and then released into constant conditions, a clear-cut tidal rhythm has been reinstilled. It is interesting, because while the imposed pressure cycle had peaks of identical amplitude, the rhythm assumed has a semidiurnal inequality with successive peaks alternating between tall and less so. Periodogram analysis indicates significant peaks at 12 h and 24 h, and the authors (Abelló *et al.*, 1991) state: ". . . data clearly show highly significant periodicities of both circatidal and *circadian* rhythmicity in activity. . . ." If that conclusion is correct, then here again is an example of a pressure cycle with a 12.4-h period initiating a circadian rhythm! This is unheard of: I think that is not the case, the periodogram peak at 24 h is either simply a supermultiple of 12 (see Chapter 2 where the penchant of periodogram analysis to produce supermultiples is discussed in detail), or the semidiurnal inequality of the rhythm is interpreted as a 24-h period by the technique.

The main prey of *L. holsatus* at the collecting site is the brown shrimp *Crangon crangon*. That animal emerges from the sand around the times of high tide and reburies at low tide (Al-Adhub & Naylor, 1975). The similarity in phase between its rhythm and that of *L. holsatus* suggests that food (i.e., the shrimp) may be a rhythm initiator and phase setter of the activity rhythm of the crab.

The Rocky Shore Crab *(Hemigrapsus)* and its Phase-Response Curve

I will start here with a definition of a so-called phase-response curve and describe its construction. For cyclic environmental factors to entrain a rhythm to their frequency, the clock must, of course, be sensitive to them. And it turns out that this sensitivity changes in a regular way during each cycle, at some times phase advances are produced while at other times phase delays result. The changing sensitivity is itself rhythmic, and is described by a **phase-response curve**: a plot of observed phase shifts as a function of the phase at which the stimulus is administered. This was first discovered in the study of circadian rhythms, and is best exemplified by them. Using Fig. 4-14, I will describe to those needing it, how one is created.

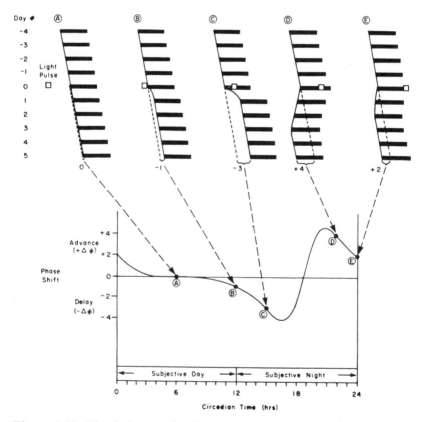

Figure 4-14 The derivation of a phase-response curve for the circadian rhythm of the likeable, but deadly predacious nocturnal animal, the *Horn swoggle*. (A) through (E) are separate experiments where identical 1-h light pulses (the open squares) are offered at different times during the animal's rhythm (the dark bars indicated when the animal was active). The different degrees, and direction, of phase change produced are plotted in the panel below, creating a response curve. See text for details (Moore-Ede *et al.*, 1982).

The figure shows the activity rhythm of, say, a single *Horn swoggle* (a resident of Jurassic Park), that in constant conditions, describes a 25-h period (only the active portion of the animal's rhythm is shown [as dark bars], and these bouts drift to the right side of the figure because the period is longer than 24 h). In (A) on the fifth day in constant conditions, a 1-h light pulse is offered at a time when the monster is asleep (and herbivores are thus resting unthreatened). That pulse causes no phase change, and this is indicated by the point on the zero line of the panel below. The experiment is repeated (B), but this time the light pulse is given just before the onset of activity, say at dusk. A small phase delay is caused, the amount of which is plotted under the zero line. A pulse at (C) causes an even greater delay, but giving the same pulse at (D) and (E) causes phase advances. In practice, more points are usually defined — but you get the idea by now. The line connecting the points is the phase-response curve that describes the shape of an underlying rhythm in changing sensitivity to light pulses. The shape of the curve indicates the adaptive nature of its existence: this is the way an organism's rhythm is properly adjusted to the day/night cycle. Suppose that in nature, *H. swoggle* should accidentally begin its active phase before darkness sets in. The pre-sunset light enters the animal's eyes and causes, as you can see from the phase-response curve, a phase delay. The next day the animal begins its active phase at the proper time, after the sun sets. Or, on the other hand, if the animal remains active after sunrise, then light falls on the phase advance part of the curve, moving the whole rhythm back into the hours of darkness.

The actual shape of a phase-response curve is a function of the duration, the intensity, and the time that a pulse is given.

Light is the prime phase setter for circadian rhythms, but, of course, does not reset tidal rhythms. Temperature, pressure, salinity, and turbulence do, and because they are able to entrain, and if the tide-associated clock mechanism is like the one controlling circadian rhythms, then it should be possible to identify an underlying phase- response curve for each of the four **zeitgeber**. The search for them will be described in the following paragraphs.

Hemigrapsus edwardsi lives, conveniently, under rocks on the plinth of the Portobello Marine Laboratory on the South Island of New Zealand. It has an interesting clock-controlled locomotor rhythm (Williams, 1969). It became the subject of an extensive study designed to see how its rhythm would respond to 10°C cold pulses. Freshly collected crabs were assigned to groups of five animals and their rhythms first recorded for 12–24 h in constant dim red light and 15°C. Then, systematically, through the interval between one low tide and the next, five-crab groups were dunked in 10°C seawater for 3 h and then returned to the previous constant conditions and their rhythm followed for 36–48 h. This experimental design tested the species for a changing phase responsiveness.

The reader is forewarned that most of the phase-response curves for tide-associated rhythms are ragged, somewhat unconvincing things, and

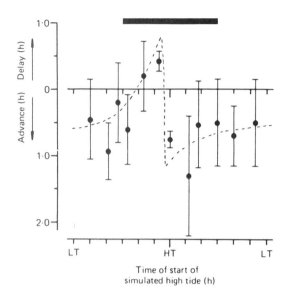

Figure 4-15 The phase-response curve for the crab *Hemigrapsus edwardsi*, created by subjecting 39 five-crab groups to 3-h cold-pulses at various times throughout one tidal cycle, and then returning them to constant darkness and 15°C. Each mean is skewered by ± one standard deviation. The bar overhead encompasses the interval of one expected high tide (Naylor & Williams, 1984).

that is the case for the *Hemigrapsus* one. The phase changes, advances and delays, for 39 separate five-crab groups are portrayed in Fig. 4-15. As indicated by the size of the standard deviation skewers on each mean, the responses were widely variable. In most cases the 3-h cold pulse caused small phase advances, but as is characteristic of most phase-response curves, at one point there is a sudden, precipitous phase change from one sign to the other. That occurred here at the expected time of high tide. Thus if the cold pulse truly represents a flood tide, then giving it before a scheduled high tide would "warn" a crab's clock that it was phased too early and signal it to delay. Pulses offered after high tide should produce phase advances (Naylor & Williams, 1984). One can almost see that in this curve. With 20:20 hindsight, a much more impressive curve might have been expected if individual crabs with precise rhythms had been specifically chosen for use, but that would have required an enormous amount of work.

There is every reason to expect that the phase responsiveness of an intertidal organism would not be anywhere near as impressive as the light-pulse-derived phase-response curves supporting circadian rhythms. In the latter case, the times of sunrise and sunset are very clear demarcations that change their timing only very slightly on a day-to-day basis. On the other hand, the tides — as described in the first chapter — are highly erratic, and are notorious for being mis-scheduled by aperiodic weather changes. It would not be adaptive for an organismic tidal rhythm to rephase to each of these sometimes large vagaries.

An aside: in Fig. 4-15, the delays are plotted above the axis and the advances are plotted below, the opposite of scientific convention and that used in modern studies of circadian rhythms. But, reader, be tolerant, this is just an idiosyncrasy of some of us who work with tidal rhythms.

The Isopod (*Excirolana chiltoni*)

This animal lives on the shores of California where it buries itself in the sand during low tides, but emerges and swims when the flood tide covers its habitat. The tides there are a mixed, semidiurnal unequal type: one peak usually being greater than the other. During each fortnight, the largest tidal peak gradually decreases in amplitude to a point of extinction, at which time the other peak, which has now grown in size, is the only one expressed during the lunar day. The extinguished peak then appears again and grows in amplitude at the expense of the other tide. During this alternation in size, a condition is reached when both tides are, for a moment, equal in depth. The end result is that on a few days each month there is only one high tide/lunar day, and on other days, two. And during the two tides/day intervals, sometimes the highest tide comes during daylight and at other times it occurs at night. Or, on some days, both tides are equal in amplitude. Because *Excirolana* lives so high on the shoreline, the low-amplitude tides on some days each month never reach the animal's subterranean habitat.

The isopod's swimming rhythm will persist in the laboratory, where its form mimics the last tides to which it was exposed at the time of collection. Thus, the rhythm may exhibit only one peak/day, or, when two peaks/day are displayed, the one corresponding to daylight may be either large or small, or equal in amplitude to the other one (Klapow, 1972). Complicated? Certainly not to *Excirolana*.

Clearly the tides are involved with determining the changing forms of this protean rhythm. In the thought that the pounding surf of the flood tide was the causal agent, both flask shakers and magnetic stirrers were exploited as "wave simulators." Isopods were allowed to become arrhythmic by storage in the laboratory, and were then exposed to 12.4-h cycles of agitation and quiescence. When released into constant conditions it was found that the rhythm had been re-initiated and the phase set by the treatment (Klapow, 1972). Or, when wave-simulator cycles were given out of phase with the rhythms of freshly collected animals they caused a phase change (Enright, 1965). Even more interesting, the form of the rhythm produced could be designed by altering the length of the agitation phase of each cycle. In this demonstration, freshly collected animals were divided into two groups of 115 animals in each. One group was subjected to 7 days of periodic agitation consisting of a 30-min duration pulse each afternoon and a 120-min one each morning; the onset of each treatment

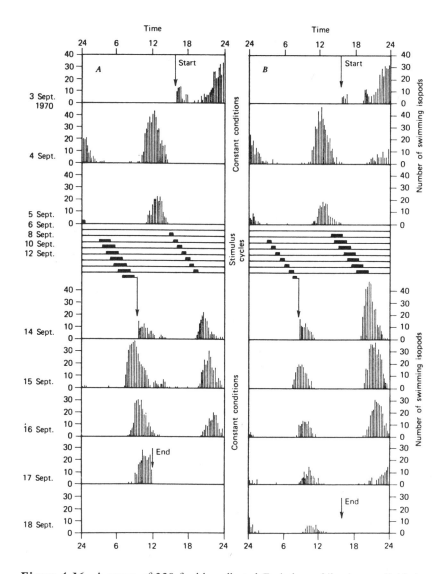

Figure 4-16 A group of 230 freshly collected *Excirolana chiltoni* were divided in half and placed in constant conditions for 2.5 days. Then they were subjected to tidally spaced intervals of "wave simulation" for 7 days. Group A received 120 minutes of agitation each morning and 30 minutes each afternoon. Group B received the same with the agitation lengths reversed. The dark bars represent the intervals of wave simulation. When the subjects were then transferred back to constant conditions, it is seen that the treatments set the phase of the rhythms and also established the inequality of the amplitude (Klapow, 1972).

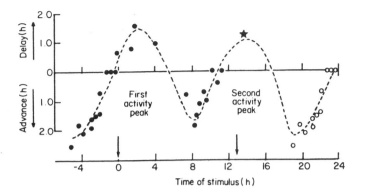

Figure 4-17 The phase-response curve to "wave simulation" for the isopod *Excirolana chiltoni*. The points above the zero axis represent phase delays (size given on the ordinate), and those below phase advances. The open circle on the right-hand side of the figure are repeats of the same points on the left. The star is the mean of several experiments using single animals (modified from Enright, 1976a).

was separated by 12.4 h (Fig. 4-16A). The other group was treated the same way except that the long period of agitation was offered in the afternoon (Fig. 4-16B). After 7 days of this treatment the isopods were switched to constant conditions. Note that the short entraining pulses produced relatively smaller peaks than the long ones (Klapow, 1972).

All of the above results were collected by studying the group responses of many individuals. It turns out that the rhythms of single animals could also be examined successfully, and like all intertidal organisms, with this species there is also a great deal of variability between subjects — including a minuscule group of "virtuosos." When the time came to tackle the labor-intensive task of defining a phase-response curve for agitation pulses, both individuals and groups of *Excirolana* were used. The results from both were approximately the same. The rhythms of freshly collected specimens were first examined in constant conditions for 3–4 days. Then each individual, or group, was subjected, at a variety of different phases throughout one activity cycle, to 10 s of shaking each minute for 2 h, and then returned to constant conditions for up to 9 days where the persistent rhythm was scrutinized for changes in phase. The effort was worthwhile: the first phase-response curve ever produced for a tide-associated rhythm was the result (Enright, 1976a,b). The masterpiece is seen in Fig. 4-17, where the falling arrows indicate the times of the expected maxima in swimming activity. Wave simulation given at the time of peak activity produced no phase change, while those given before then caused phase advances, and those after that point, phase delays.

Enright interpreted the *Excirolana* cyclic expression as ". . . a tidally synchronized bimodal circadian rhythm. . . ." Believing this, after determining the phase responses of the first peak in activity, he went on to determine them for the second same-day maximum. This produced the only bimodal phase-response curve ever described for a circadian rhythm.

In the many years that have passed since the *Excirolana* phase-response study was completed, I have discussed (e-mail is wonderful: world-wide consultation reduced to a keyboard and microseconds) the above conclusion with many other marine chronobiologists and have encountered none that agrees with the *bimodal circadian* interpretation. The second hump of the so-called bimodal pattern is explained by most as simply the peak of the next tide-associated activity cycle.

Actually, there are several features of the *Excirolana* rhythm — discovered during wave-simulation experiments — that support the circalunidian clock-control hypothesis. The fact that one or the other of the two peaks described in the laboratory each lunar day can disappear and later reappear without involving the other, exemplifies their independence. If arrhythmic *Excirolana* are exposed to just one agitation pulse, just one activity peak forms (Klapow, 1972), meaning, according to the hypothesis, that just one of the two circalunidian clocks had been activated. And in the phase-testing experiments of Enright (1976b), when he offered an agitation pulse timed several hours after a major peak in a *unimodal* activity-rhythm display, in addition to phase shifting the existing peak, it often also caused the initiation of a new peak (Fig. 4-18). And that newly induced peak always arose at the same time, c.12.4 h after the previous peak, in spite of the fact that the agitation stimulus could be given anytime during a 12-h window centered over the point where the new peak appeared. That is the kind of result that would have been predicted by the circalunidian-clock hypothesis: The two lunar-day clocks are coupled together antiphase, so that when an agitation pulse "awakened" the second clock causing it to produce a second locomotor-activity peak, that peak appeared, because of the nexus between the two clocks, half-way between the other clock's successive peaks, rather that at the exact point at which the shaking pulse had been given.

The Isopod *(Eurydice pulchra)*

Jones and Naylor (1970) tested a variety of possible phase-setting stimuli on the *Eurydice* swimming rhythm. Twelve-hour light intervals alternating with 12-h dark pulses would not entrain, but 12-h cycles of hydrostatic pressure change were effective. The regimen consisted of cyclic, 30-s pressure fluctuations equivalent to 5 m of water, offered for 30 min, alternating with 11.5 h at sea-level pressure. These cycles were offered for 5 days and then the animals were released into constant conditions where the pattern of their swimming was examined. A 12-h cycle was displayed with peak swimming coming at the times of expected high pressure.

Agitation cycles were also found to be effective. Using a laboratory

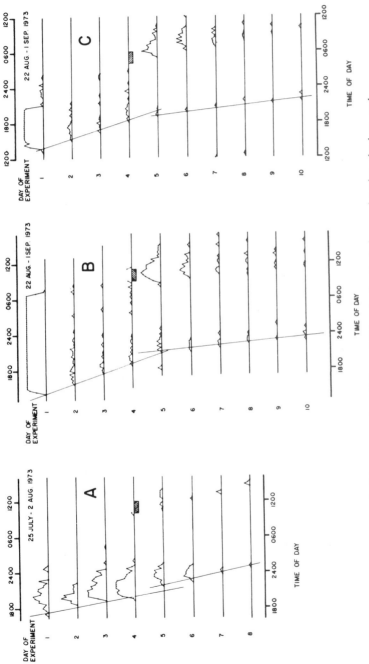

Figure 4-18 Three examples of phase shifting in the *Excirolana* swimming rhythm, and also initiating a second string of activity bouts by a *single* 2-h "wave-simulation" offered at the times indicated by the shaded bars (Enright, 1976b).

flask shaker, arrhythmic animals collected at spring tides were shaken for 30 min at 12-h intervals. The treatment was repeated for 4 days and the animals studied in constant conditions. The treatment established a persistent rhythm with peaks being expressed at the times of expected shaking (Hastings, 1981).

Temperature entrainment has not been tried on the *Eurydice* swimming rhythm. However, rephasing with single cold-temperature pulses has been attempted, and found to work. Arrhythmic animals were subjected to a 3-h pulse of 5°C and returned to 18°C seawater. The burst of activity expressed after the pulse was then repeated at 12.4-h intervals thereafter (Jones & Naylor, 1970).

The Amphipod (Corophium volutator)

In Chapter 3, it was described that re-establishment of the activity rhythm in *Corophium* could be bought about by returning arrhythmic animals, incarcerated in a porous container, to the intertidal zone for 3 days. Just which aspects of the tide were active ingredients in the re-establishment was unknown. Therefore the influence of periodic inundation *alone* was the first of several influences tested in the laboratory. One hundred animals were placed in an aquarium half filled with sand and water; the water existed only in the interstices between sand grains, and reached a level 5 cm below the surface of the sand — well beneath the bottom of the animals' shallow U-shaped burrows. Water and sand were held at the same temperature (*c*.15°C). Then, during the last 60 min of consecutive 12.5-h intervals, the water level was brought up to the sand surface and drained again. It was found that the amphipods only came to the entrance of their burrows during these short "high tides" (the wetting interval was 4 h shorter than the normal duration that the animals would have experienced in nature). Over an 11-day interval, 21 repetitions of the inundation cycle were offered antiphase to the times of the expected normal tides. Then the animals were sieved out of the sand, placed in seawater, and their swimming activity monitored. It might be postulated that if entrained their activity peaks would be centered at the times of the expected (1 h) high tide pretreatment. But they were not; it was the sharp minima of their rhythm that were so re-aligned (Fig. 4-19)!

The experiment was repeated, but this time the wetting interval was lengthened to 6 h, slightly longer than the natural interval. After 11 days of exposure these animals were released into constant conditions where their rhythm was found to have been entrained to the time of high water (Holström & Morgan, 1983a).

Various other versions of immersion cycles were also tested. The basic cycle length was 12.4 h with immersion lengths ranging from 2.5 h to 10 h. These cycles were repeated for approximately 8 weeks, after which the animals were released into constant conditions and their swimming pattern recorded. The basic *c*.12.4 h period persisted in all cases.

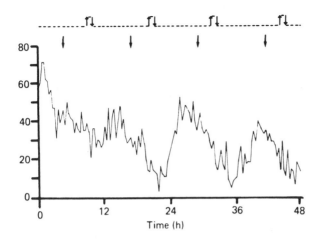

Figure 4-19 The group swimming rhythm of *Corophium volutator*. The animals were first exposed to 21 cycles of 1-h immersions alternating with 11.5-h stints when the water was withdrawn; and were then suspended in 15°C seawater. The up and down arrows at the top of the figure indicate the subjective times of 1-h inundations. Note that, unexpectedly, it was the valleys of the rhythm that were entrained to the times of flooding. The lower row of arrows indicates the times of expected high tide in nature (Holström & Morgan, 1983a).

Next the experiment was repeated but the cycle length was varied: periods of 4 h, 6 h, and 8 h were tested. After 10 weeks of exposure to 8-h cycles with 5-h inundation stints, the animals showed *c.*12.6 h periods in constant conditions. After 10 weeks of 6-h cycles with 4-h inundation phases, peaks in constant conditions tended to occur at 6-h intervals, however the first and third peaks were of large amplitude and separated by 12 h, suggesting an attempt to reveal a tidal period. Four-hour cycles with 2-h immersions did not entrain.

The last version of observations used 24-h cycles, with 6-h, 8-h, 10-h, and 12-h immersion phases. The last three produced significant swimming rhythms of 11.9 h, 12.6 h, and 12.2 h (Harris & Morgan, 1984a).

Corophium is quite tolerant of low temperatures, and this property enabled the following study that attempted to rephase its swimming rhythm with cold pulses. Temperatures as low as zero were ineffective, but −1.5°C worked with summer-collected animals, while a chilly −4.5°C was required for those collected in the winter. Freshly captured animals were held in the laboratory at 15 ± 1°C for *c.*18 h and then subjected to a 2-h or 3-h sub-zero cold pulse. The procedure seems drastic — and certainly it was not a natural stimulus for the summer animals — but survival was plainly adequate to extend the investigators' publication record by one. In separate experiments pulses were given at a different stage of an expected tidal cycle. The results are seen in Fig. 4-20: pulses applied during the flood

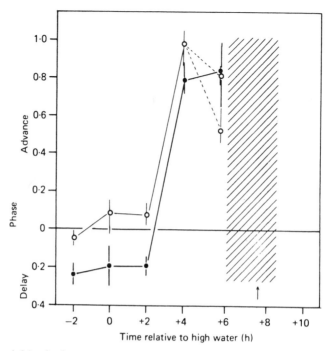

Figure 4-20   A phase-response curve constructed for *Corophium volutator*, by subjecting groups of animals to 2-h and 3-h (the open and closed circles, respectively) −1.5°C temperature pulses at the times indicated on the abscissa. "The shaded area indicates a period of equivocation during which cold pulses of both 2-h and 3-h duration induced arrhythmic swimming." The vertical bars on each point are not identified, nor is any mention made of the significant differences between points. The two open circles over hour 6 indicate that the peak split after chilling at this time (Holström & Morgan, 1983b).

tide produced slight delays in the expression of the activity peak, while pulses given during the ebb caused small advances. Pulses applied during the times of expected low water rendered *Corophium* arrhythmic. Because only sub-zero temperatures were effective, it is impossible to assign any *natural* use to this finding, but it does show that orderly internal changes, probably clock controlled, are taking place within the organism (Holström & Morgan, 1983b; Morgan, 1991).

A quick study of the temperature entrainability of the activity rhythm of *Corophium* has also been undertaken. Freshly collected animals were subjected to 12.5-h, high/low temperature cycles. Thirteen cycles of 6.5-h stints at 15°C, alternating with 6-h reductions in temperature were tried; each change-over consisted of a gradual 3-h drop to 5°C, followed by a 3-h rise back to 15°C (Fig. 4-21). This cycle did not mime the one that occurs in nature where flooding is rapid; the advancing front applies a sudden temperature change as it sweeps over an amphipod's burrow

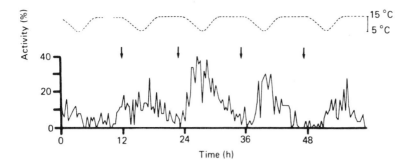

Figure 4-21 The activity rhythm of *Corophium* in constant light and temperature (15°C) after being exposed to 5 days of a 5°C–15°C temperature cycle (whose expected phase appears above the curve (Holström & Morgan, 1983b).

opening. The temperature cycle was applied so that the interludes of low temperature corresponded to the times of expected high tide. Because of the jagged peaks of this rhythm it is difficult to assign just how the maxima are juxtaposed with the temperature cycle, but it is fairly safe to say they corresponded pretty well to the times when the temperature cycle began to decline from 15°C to 5°C. At the end of this exposure, the animals had been in the laboratory for about 9 days, meaning that the would-be control animals had now become arrhythmic, so that there was nothing to compare the experimentals to when they were released into a constant 11°C. The first maximum exhibited was such a broad saw-tooth of peaklets that it is really impossible to describe its phase (unless one is willing to pretend that one point makes a peak), but the next two peaks clearly fell at the times of expected temperature minima. Therefore, it looks like entrainment had taken place (Fig. 4-21). One thing that is certain, is that offering a temperature cycle kept the rhythm from damping out (Holström & Morgan, 1983a).

The estuarine environment in which *Corophium* lives is subject to periodic changes in salinity associated with tidal exchange. *Corophium* is a good osmoregulator that can tolerate salt concentrations in its ambience between 2‰ and 47‰. With this built-in ability, the animal was a likely candidate for phase setting and entrainment studies using salinity as a zeitgeber.

In phase setting with 40‰ seawater, single pulses of 3 h, 6 h, and 9 h durations were found to be effective. I will give one example. Four groups of 10 individuals, all suspended in 30‰ seawater at 12 ± 1°C, were tested simultaneously and the results seen in Fig. 4-22. Group A received no pulse and served as a control. Groups B, C, and D, starting at the moment of the predicted high tide, received 3-h, 6-h, and 9-h pulses of 40‰ artificial seawater held at a temperature of 12 ± 1°C. As can be seen, all pulses caused a delay (the relationship being approximately logarithmic), the 9-h one

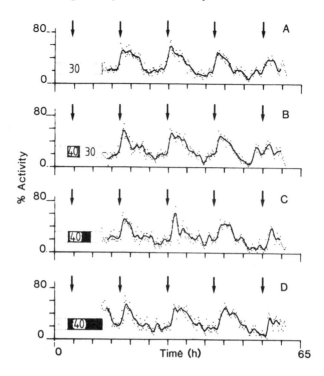

Figure 4-22 The activity rhythm of *Corophium* at a temperature of $12 \pm 1°C$ in 30‰ seawater. The points represent a short interval of swimming activity, and the curve drawn through them is a smoothing created by a 15-h moving average. The falling arrows indicate the times of subjective high tides. Beginning at a high tide, 3-h, 6-h, or 9-h pulses of 40‰ seawater (B, C, & D, respectively) were offered. Each pulse is indicated by the number 40 in an otherwise black bar; at the end of each pulse the animals were released back into 30% seawater. A is a 30‰ control. The results are analyzed in Fig. 4.23 (Harris & Morgan, 1984b).

producing the largest, a 90 min lag. The regressional relationship of delay relative to the controls is seen in Fig. 4-23; the association is significant: $r = 0.763, P = 0.001$).

Next, the rephasing effect was tested as a function of the phase in the swimming cycle that it was offered. The standard pulse used was 3 h of 40‰ seawater, which was begun 1-h later in the cycle with each experiment. Figure 4-24 shows the result: a phase-response curve suggesting a regular pattern of phase advances and delays. The units of the phase changes produced are not given — I suppose they are tenths of hours. If that is true, the changes are slight.

To test the entraining ability of salinity cycles, 3-h intervals of 10 or 40‰ seawater were alternated with 9.5-h stints of exposure to 30‰. No influence was found when using 10‰. In the test of 40‰ seawater pulses,

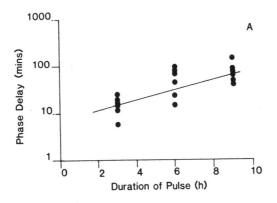

Figure 4-23   A regressional analysis of the data seen in Fig. 4.22: delay versus pulse duration. Each point represents the phase delay created by a 40‰ pulse given for the duration indicated on the abscissa, relative to the phase of the rhythm recorded in 30‰ seawater. Their relationship is approximately logarithmic (r = 0.763; *P* = 0.001) (Harris & Morgan, 1984b).

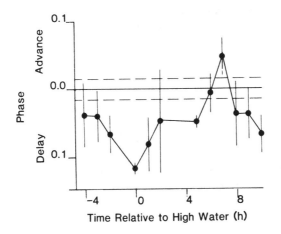

Figure 4-24   A phase-response curve based on the results of subjecting the *Corophium* persistent swimming rhythm to 3-h pulses of 40‰ seawater administered at the times indicated on the abscissa. The first three points have been repeated for clarity. The vertical brochettes through each point are not identified, nor is the unit of phase change (I suppose they are standard errors, and tenths of hours, respectively). The solid horizontal line is the mean, and the broken line ± one standard deviation, of the phase of the rhythm relative to the tide while exposed to 30‰ seawater (Harris & Morgan, 1984b).

freshly collected individuals were used, and curiously, the pulses were offered in phase with the expected high tides. Thus, during the 3 times that high salinity was imposed, the phase of the experimentals looked to be identical (to me) to the controls. And after the release of the experimentals into a constant 40‰ (again to me, looking at the published figure) that relationship seemed still to be obtained. But the authors say that a small difference existed — and after all, they did the statistics (which are not given) and should know what really happened (Harris & Morgan, 1984b).

Morgan (1965) subjected this animal to sinusoidal pressure-change cycles matching the interval of the tide. During the treatment the animals overtly rephased to the times of high pressure. Unfortunately, Morgan did not then test the apparent entrainment in constant conditions.

The Amphipod *(Synchelidium sp.)*

Another amphipod, a co-inhabitant with *Excirolana*, displays the same persistent rhythm as does the isopod. Populations of *Synchelidium* were placed in soft flexible bottles and anchored on the intertidal sea bottom. Because of the deformable walls of the container, tidal pressure changes outside were impressed on the incarcerated animals. But, after bottles were retrieved (doubly important nowadays for the recycle refund) the animals' activity rhythm did not reflect the phase of the tides to which they had been exposed. A sinusoidal 12-h pressure cycle that mimicked the pressure changes in the animals' habitat, but that was offered to captives in the laboratory, was also an ineffective zeitgeber (Enright, 1962; 1963).

The Amphipod *(Bathyporeia pelagica)*

The tidal swimming pattern of this little animal will persist in constant conditions for 5 days, with the main activity bursts roughly synchronized to the times of the expected early ebb of the tides. In one brief experiment it was found that its rhythm could be rephased by a very cold pulse. A single 1°C, 6-h pulse, centered on the peak of the swimming rhythm, was found to delay the animal's rhythms by 3 h and 20 min (Fincham, 1970).

Mollusks

The role of very cold temperature pulses has been tested on the New Zealand clockle *Austrovenus stutchburyi*. Fourteen clams were subjected to an 18.5-h stint in 2°C seawater. At the end of the pulse the animals were returned to 14.5°C where they immediately opened their shells in spite of the fact that it was a time when they would have been expected to keep their valves closed. A typical example is seen in Fig. 4-25; the return set a new phase that held thereafter (Williams *et al.*, 1993).

The next two experiments had a dual purpose: (1) to test the role of inundation cycles as a zeitgeber; and (2) to see if it is possible to entrain just one of the two lunidian clocks postulated (Chapter 3) to control the gaping rhythm in this shellfish. After animals had been stored in the laboratory for over two months, their gaping was tested for several days to ensure arrhythmia (Fig. 4-26, top). They were then returned to the

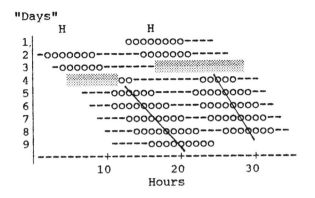

Figure 4-25 The effect of an 18.5 h pulse of 2°C seawater (stippled bar) on the shell-gaping rhythm of *Austrovenus stutchburyi*. When the clockle was returned to 14.5°C water it is seen that low-temperature treatment produced a large phase delay. Note the difference in period lengths: one was 25.7 h, and the other 25 h. O = shell open. H = high tide on day of collection (modified from Williams *et al.*, 1993).

intertidal and exposed to one daytime tidal cycle (one low, and one high, tide). Then they were returned to the laboratory where they were kept in air and dim light until the next day when they were returned to the shore for one more tidal cycle. Thus, these clams had been exposed to only two high tides during an interval of 50 h. When their shell gaping was then studied in constant conditions, with the exception of one small peak on the first day coming at the "wrong" time, they opened only at the times of expected daytime high tides (Fig. 4-26, bottom). As described in Chapter 3, the clockle lacks a solar-day rhythm, so the unimodal display must represent that of a tide-associated rhythm. Because only one tidal peak was initiated, these data certainly give strong support to the circalunidian clock hypothesis.

So good so far. The experiment was repeated with only one slight change: the clams were exposed to two night-time, only, tidal cycles rather than daytime ones. When shell gaping was then tested in constant conditions, to everyone's surprise, both the daytime and night-time tidal peaks were exhibited (Fig. 4-27 bottom)! Maddening; but why the difference? As of this writing, these are pristine experiments that must be repeated and varied before the truth be known (Williams, 1995). It could be that the conditions in the second experiment permitted a re-coupling of the two lunidian clocks; thus, starting one, automatically started the other. Time will tell (no pun intended).

The Shanny *(Lipophrys pholis)*

As might be expected, the first approach to learning how fish rhythms are entrained by the environment was to expose arrhythmic fish to a certain

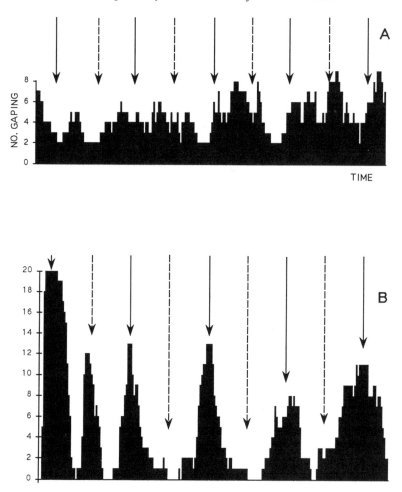

Figure 4-26 The role of inundation as a phase setting stimulus for the clockle shell-gaping rhythm. *Top:* Shell gaping in 20 clams after more than 2 months in the laboratory. There is no indication of a group tidal rhythm. *Bottom:* After two daytime (only) exposures to the tide, the animals exhibited a tide-associated rhythm in constant conditions in which they gaped only (except for the small peak on the first day) at expected daytime high tides. Number of clams gaping on ordinate, and days on the abscissa. Falling arrows indicate the times of expected high tides, the ones with the solid shafts being the daytime high tides (Williams, 1995).

number of tidal exchanges to see what happened. Gibson (1971) placed caged shannies that had become arrhythmic during laboratory storage, out on the shoreline and found that just a few exposures to the tides would induce them to become rhythmic again. Transplanting caged fish to a shoreline with a different tidal schedule caused a phase change to the new locale. Offering cycles of temperature change, feeding, or turbulence did

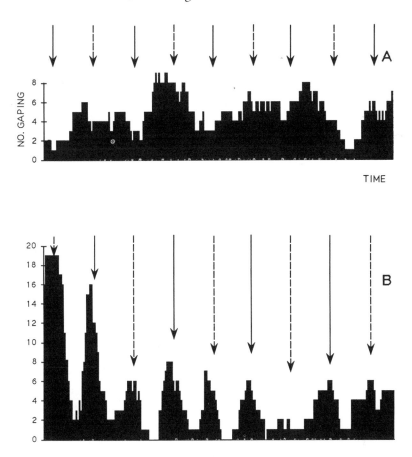

Figure 4-27 The role of inundation as a phase-setting stimulus for the clockle shell-gaping rhythm. *Top:* Shell gaping in 20 clams after more than 2 months in the laboratory. There is no indication of a group tidal rhythm. *Bottom:* After two night-time (only) exposures to the tide, the animals exhibited a tide-associated rhythm in constant conditions in which they gaped at both the expected daytime and night-time high tides. Number of clams gaping on ordinate and days on the abscissa. Falling arrows indicate the times of expected high tides, the solid-shafted ones in the lower data set being the night-time high tide (Williams, 1995).

not reinstitute the rhythm. Also, suspending arrhythmic fish just under the ocean surface in a floating cage did not restart their rhythms; comparing this outcome with the result of caged fish on the ocean floor suggests a good chance that hydrostatic pressure is a zeitgeber.

A confirmation of some of the above, in a study that also added much new information, has been provided by Northcott *et al.* (1991a). The study consisted of variations of the above approach of exposing arrhythmic caged fish to 1–4 natural tides. The first parameter to be tested was the influence of the cage itself. Freshly collected fish were caged, exposed to 4 tidal exchanges on their native shore, and then placed in constant conditions. At the same moment they entered constant conditions, another group of fish, freshly collected, were treated the same way. Unless the cage played some detrimental role, the number of rhythmic fish in both groups should be about the same. But only 47% of the former were still rhythmic, while 87% of the latter remained so, and that difference was statistically significant. Thus, caging per se appears to reduce the entraining influence of the tides.

There are several possible explanations for cage-reduced synchronization ability. It may be that the cage somehow reduces the "strength" of the environmental phase setters, but it is hard to believe a cage made of simple mesh could do that. Possibly there is a reduction of food supply that leads to starvation and stress, but animals can remain rhythmic in the laboratory for many more days without being fed. A likely cause could be that the small cages reduced the amount of activity, and feedback from the reduction somehow disables the coupler or the clock (a case where the driver is driven somewhat by an environmental driver). The idea is re-enforced by the results of Green (1971a,b) whose study was not plagued by this difficulty because he did not have to cage his animals when they were released; they have a strong homing instinct and returned to their home pool where they could be rather easily recaptured. Their rhythmicity had not been reduced when tested after recapture.

From the above it should be clear that one cannot expect to find that all arrhythmic shannies have become rhythmic after exposure in cages to the tides. In fact, only about one third of 78 fish did entrain to 1–4 tidal exposures. A positive example of one such experiment, an exposure to two tides, is seen in Fig. 4-28.

The last variation was a test of the circalunidian hypothesis for the clock control of tide-associated rhythms. A special cage was built that would retain enough water in its base for the captured shannies to remain covered when the cage was lifted out of the sea. Five arrhythmic animals were placed in it and exposed to one tidal inundation; the cage was then lifted up out of the water, carried inshore, and placed on dry land where the next high tide would not reach it. Then, after that tide had come and gone, the cage was returned to the lower position on the shore to be inundated by the next tide. The next trip was to constant conditions where swimming was monitored. (A shorter version of the same treatment was carried out when fish were exposed to only one tidal dunking.) The question to be answered was this: If there are two "circalunidian" clocks, one for each of the two daily peaks, will the exposure to only every other tide start only one clock (or re-couple it with the rhythm it drives)? If only one peak was then displayed at $c.24.8$-h intervals, the hypothesis would have been supported, but if both peaks appeared it would not be clear what happened: it could mean that there was only a single clock running at $c.12.4$ h, or because of the strong coupling (it must be tenacious since it

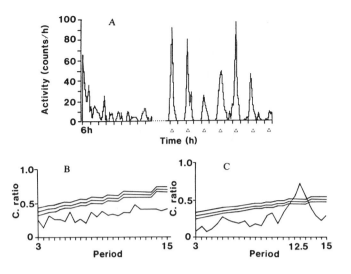

Figure 4-28   The role of natural tidal inundation on the swimming response of a single *L. pholis*. (A) Swimming before and after an exposure to two natural tidal cycles; the dotted line indicates caged time, and the open triangles the time of scheduled tides on the shoreline. (B) A periodogram of the activity pattern shown on the left side of (A). The three lines above the periodogram represent the 0.05, 0.01, and 0.001 levels of significance. Clearly, there is not a significant cycle in those data. (C) A periodogram of the data after the fish had been exposed to the tides showing a strong significant peak at 12.5 h (Northcott *et al.*, 1991a).

breaks spontaneously only rarely) between the two circalunidian-clock-driven peaks, activating one might be expected to drag along the other one.

The results of the experiment (Northcott *et al.*, 1991a) were a tease of both possibilities: all five fish showed the typical two peaks/lunar-day persistent rhythms, but in one fish, one of the peaks was more than double the amplitude of the other (Fig. 4-29) (Northcott *et al.*, 1991a). And the great peak was the one that corresponded to periods when the fish had previously been inundated by the tide! Is that a subtle clue that we are seeing two clocks at work in this shanny?

Hydrostatic pressure has been found to be one of the most effective phase-setting and entraining agents for the rhythms of intertidal fish. The apparatus created to produce these pressure changes, as well as the activity recorder, are seen in Fig. 4-30. Brilliant in its simplicity, the apparatus consists of a constant-head reservoir mounted on a rotating vertical arm. The higher the reservoir is raised the greater the pressure, and the rate that the supporting arm is rotated determines the period of the pressure cycle produced (Gibson, 1982). The device is especially nice in that it produces gradual (rather than square wave) pressure changes — as would the

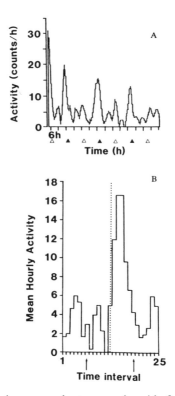

Figure 4-29 After being exposed to every other tide for three cycles, this shanny produced the persistent rhythm seen in (A). The solid subtending triangles indicate the times of expected tide to which the fish had been previously subjected. (B) Form estimate of the same data aligned to 25 h. The peak to the right of the vertical dotted line is over the times corresponding to the previous exposure to the high tides (Northcott *et al.*, 1991a).

migrating tides on a calm day. An example of its effectiveness is seen in Fig. 4-31, where there can be no question of its worth in producing waveforms in both pressure and animal locomotion (Graham *et al.*, 1987). It is rather curious that the cycles would be so effective with *L. pholis* because it lacks a swimbladder, the organ usually thought to be the pressure receptor of fishes.

In one series of experiments, a basic 12.4-h cycle, constructed of a stint at atmospheric pressure alternating gradually with 1.3 atmospheres (equal to 3 m of water overhead, and comparable to that experienced on the lower shore of this fish's habitat during spring tides), was given first to animals that had become arrhythmic after storage in constant conditions, and second to freshly collected (i.e., hopefully rhythmic) fish. With the former it was an attempt to re-initiate cyclic behavior; with the latter it was a test of whether the pressure treatment would entrain the rhythms to its

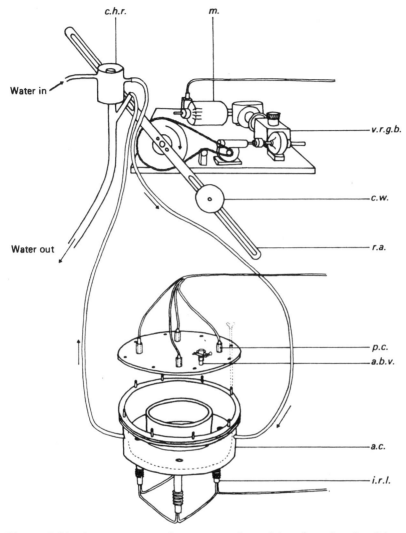

Figure 4-30 An apparatus used to measure the activity of a swimming fish, and to also subject the animal to hydrostatic pressure cycles. a.b.v., air bleed valve; a.c., activity chamber; c.h.r., constant head reservoir; c.w., counter weight; i.r.l., infra-red light source; m, motor; p.c., photocell; r.a., rotating arm; v.r.g.b., variable-ratio gearbox. Activity was recorded automatically each time a fish in the circular raceway swam through one of the four infra-red light beams (Gibson, 1982).

frequency (i.e., prevent the rhythm from assuming a circa period). Eight to 16 such cycles were offered in a single observation.

During exposure to these cycles, 41% of the 42 arrhythmic fish increased their swimming activity during intervals of high pressure, i.e., they at least looked like they had become rhythmic. But when then tested

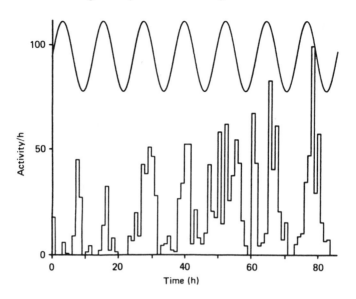

Figure 4-31   An example of the effectiveness of the apparatus pictured in Fig.
4-30 in initiating and entraining an arrhythmic shanny, *Lipophrys pholis* to
12.4-h cycles of atmospheric pressure versus atmospheric + 3 m water (Graham
et al., 1987).

in constant conditions, only eight of them showed significant tidal rhythms
— and while the mean period of this group during the attempt at
entrainment was 12.55 ± 1.75 h, the mean period of those whose rhythms
returned was a variable 11.6 ± 2.25 h. The latter mean was calculated
without including the period of one "outlier" rhythm. That fish had a
large-amplitude rhythm with a significant ($P < 0.01$) period of 24.5 h —
a circalunidian period (the species does not possess a solar-day clock) that
persisted both during the imposed pressure cycles, and subsequently in
constant conditions (Northcott *et al.*, 1991c).

The effect of this pressure cycle — phased in the laboratory to mimic
those in the natural habitat — on freshly collected fish was rather curious.
The period of the fish's rhythms averaged that of the 12.4-h entraining cycle, while the controls in constant conditions stretched out to
13.04 ± 0.85 h. The difference between the two groups was significant:
$P < 0.05$. A curious feature in the display of the fish in the pressure cycle
is that the period length of successive cycles alternated between long and
short (Fig. 4-32). Needless to say, that was, of course, accompanied by
alternating phase changes. Two circalunidian clocks, uncoupled from one
another and running at slightly different rates would produce such an overt
display, but it seems unlikely that so many individuals' clocks would
become uncoupled simultaneously (unless pressure cycles happen to have
that effect).

Next, freshly collected fish were subjected to the usual pressure cycles,

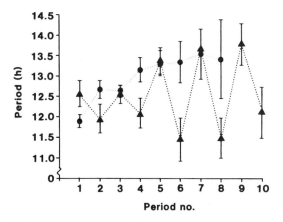

Figure 4-32 The means, decorated with ± one standard error, of consecutive period lengths of freshly collected fish (solid circles) in constant conditions; and equally fresh fish experimentals (solid triangles) exposed to precise, laboratory-generated pressure cycles. Why the latter should alternate in length is unknown (Northcott *et al.*, 1991c).

but offered 180° antiphase to the assumed rhythms, i.e., high pressure was imposed at the times of expected low tides. Surprisingly, considering the results just described above, many of the shannies ignored the cycle. Some of the fish continued to display their normal rhythm, and also described increased activity at the time of increased pressure; thus producing an overt display of one peak every 6.2 h. But none of the fish that were subsequently observed in constant conditions showed *any* rhythmic behavior!

In conclusion, these observations indicate that hydrostatic pressure cycles, characterized by gradual increases to peaks representing 3-m overhead water columns, *can* (but do not always) re-establish lost rhythmicity, and can (sometimes) entrain the *L. pholis* clock(s) to their period and phase. But paradoxically, they were totally ineffective in inverting apparent established rhythms, a 180° flip-flop being too great a challenge. And although each imposed cycle was precisely identical (machine generated), as a group they produced periods in an organismic rhythm that alternated between too long and too short (Northcott *et al.*, 1991c)! Remember that when *Carcinus* was entrained by pressure cycles, high-amplitude peaks sometimes alternated with low-amplitude ones (Fig. 4-2). These results are all parts of the conundrum those of us sufficiently masochistic to dabble in the study of tidal rhythms must tolerate.

The rock goby, *Gobius paganellus*, a teleost living side-by-side with the shanny, is also rhythmic in constant conditions (as described in Chapter 3). It too has been subjected to the same pressure-cycle treatment as *L. pholis*, and responds in the same way: some arrhythmic fish are stimulated to rhythmicity and then remain so when transferred to constant conditions, and some rhythmic fish can be entrained to the cycles (Northcott, 1991).

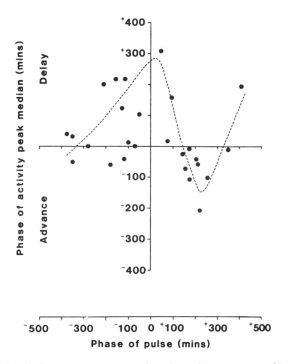

Figure 4-33 A phase-response curve, based on the responses of individual *Lipophrys pholis*, subjected, after 2 days in captivity, to a 2-h pressure pulse of 1.3 atmospheres. The degree of phase change (if any) of the next peak of the fish's rhythm is indicated by the dashed line (fit by eye) relative to the centrally located, calibrated (in minutes) ordinate. Time 0 on the abscissa is the moment of high tide, and the solid circles indicate the midpoints of the pressure pulses relative to high tide (Northcott *et al.*, 1991b).

In further experiments with the shanny, its phase responsiveness was tested. After a fish's rhythmicity was determined for two cycles in the laboratory, it was challenged at various times in its activity cycle with a 2-hour near-square-wave pulse of high pressure (1.3 atmospheres), and then returned to constant light and temperature. Because the shanny rhythms damp rather quickly in the laboratory, only the position of the first post-treatment peak was measured. A changing sensitivity to the pressure pulse was found that is represented by the response curve seen in Fig. 4-33. Maximum phase *delays* (almost 5 h in length) were found for pulses applied just after expected high tide; and maximum phase *advances* (nearly 4 h) appeared about 3 h after high tide. This result represents two *firsts*: the first phase-response curve based on *pressure pulses*, and the first tide-associated rhythm in a *vertebrate* to be so phased. An unavoidable possibility exists that the curve portrayed is not accurate in that it may be based on early transient phases, rather than on a steady-state change. (In some organisms subjected to a zeitgeber pulse, several changing periods, called transients, follow before a new steady-state phase is reached.) Until (if ever) a way can be created to keep this fish's rhythm persisting longer in the laboratory, we will not know the answer to this question (Northcott *et al.*, 1991b).

As mentioned above, Gibson (1971) had tried, unsuccessfully, to re-initiate rhythms in the shanny using turbulence cycles. Morgan's laboratory has repeated the attempt with somewhat more success. Two different types of apparatus were tried; the most successful version was a tank mounted on a teeter- totter that was tilted back and forth through a 10° arc at 10 cpm by a small motor. The swimming movements were recorded in the usual manner. Five arrhythmic fish were subjected to 11 cycles consisting of 2 h of teetering (simulated wave action) alternating with 10 h of nonagitation. During this treatment, four fish were very active during the intervals of shaking, but when they were then tested for 5.5 days in constant conditions, while it appears that four of the five fish were rhythmic for the first four cycles, periodogram analysis based on the whole data string indicated no significant rhythm. The same five fish were then subjected to 80 repetitions of the cycle, and again studied in constant conditions. Recalcitrant fish No. 5 was still non-rhythmic, and now No. 4 entered into an acyclic covenant with it; but the other three fish displayed clear *c*.12-h rhythms.

The experiment was performed one last time with a variation: the 2-h wave simulation was repeated only once every 24 hours. After a one-month rest to ensure arrhythmicity, the same five fish were used again. (What loyalty these investigators showed to old, temporal throw-away No. 5; I suppose as members of a welfare society they are so used to the doers carrying the nonproducers that it never occurred to turn him into bait or chowder.) As seen in Fig. 4-34, three of the fish (4 and 5 just swam randomly) confined most of their activity to the times of teetering. When transferred to constant conditions, for the first 4 days activity peaks of roughly equal amplitude were expressed at 12-h intervals with a phase relationship such that every other one was in synchronization to the times of the expected agitation. As usual, fish 4 and 5 (the dole fish) did not participate (Morgan & Cordiner, 1994).

Thus we are able to conclude that simulated wave action can restart and entrain the *L. pholis* swimming rhythm. Because it *appears* (one has no idea how well the teeter-totter simulates wave action) to take many repetitions to bring about an effect, this probably means that in nature this is only one component of many zeitgeber influences. When all are served up together, as in caged-exposure experiments, only a single inundation is required to initiate a rhythm. One can only guess at how offering one pulse every 24 h can bring on the desired effect: maybe that frequency is sufficient to start a *c*.12.4 h clock, or maybe it starts a lunidian clock that, via the coupling to its antiphase mate, gets the second peak running also. The changing amplitudes of the peaks of the restarted rhythm offer no clues.

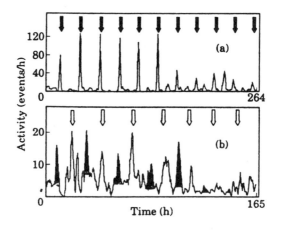

Figure 4-34 (a) The exposure of five arrhythmic *Lipophrys pholis* to 2-h wave simulations (see text for explanation). The times of exposure are indicated by the falling arrows spaced at 24-h intervals. Note that swimming activity is mainly restricted to the times of agitation. (b) After the above treatment the fish were transferred to constant conditions where peaks continued to build at the times of expected agitation (the unshaded peaks), and also half way between these maxima (the shaded peaks). The open falling arrows (spaced inaccurately) are supposed to indicate the times of expected agitation. The responses of five fish were combined to construct the curves, but two of them produced only random swimming throughout (Morgan & Cordiner, 1994).

Literature Cited

Abelló, P., Reid, D.G. and Naylor, E. 1991. Comparative locomotor activity patterns in the portunid crabs *Liocarcinus holsatus* and *L. depurator*. *J. Mar. Biol. Ass. UK*, 71: 1–10.

Al-Adhub, A.H.Y. and Naylor, E. 1975. Emergence rhythms and tidal migrations in the brown shrimp *Crangon crangon*. *J. Mar. Biol. Ass. UK*, 55: 801–810.

Atkinson, R.J. and Parsons, A.J. 1973. Seasonal patterns of migration and locomotor rhythmicity in populations of *Carcinus*. *Neth. J. Sea. Res.*, 7: 81–93.

Bolt, S.R. and Naylor, E. 1985. Interactions of endogenous and exogenous factors controlling locomotor activity rhythms in *Carcinus* exposed to tidal salinity cycles. *J. Exp. Mar. Biol. Ecol.*, 85: 47–56.

Bolt, S.R, Reid, D.G. and Naylor, E. 1989. Effects of combined temperature and salinity on the entrainment of endogenous rhythms in the shore crab *Carcinus maenas*. *Mar. Behav. Physiol.*, 14: 245–254.

Enright, J.T. 1962. Responses of an amphipod to pressure changes. *Comp. Biochem. Physiol.*, 7: 131–145.

Enright, J.T. 1963. The tidal rhythm of activity of a sand-beach amphipod. *Z. vergl. Physiol.*, 46: 276–313.

Enright, J.T. 1965. Entrainment of a tidal rhythm. *Science*, 147: 864–867.

Enright, J.T. 1976a. Re-setting a tidal clock: a phase-response curve for *Excirolana*. In: P.J. DeCoursey (Ed.), *Biological Rhythms in the Marine Environment*, pp. 103–114. University of South Carolina Press, Columbia.

Enright, J.T. 1976b. Plasticity in an isopod's clockworks: shaking shapes form and affects phase and frequency. *J. Comp. Physiol.*, 107: 13–37.

Fincham, A.A. 1970. Rhythmic behaviour of the intertidal amphipod *Bathyporeia pelagica*. *J. Mar. Biol. Ass. UK*, 50: 1057–1068.

Forward, R.B., Douglass, J.K. and Kenney, B.E. 1986. Entrainment of the larval release rhythm of the crab *Rhithropanopeus harrissii* by cycles of salinity change. *Mar. Biol.*, 90: 537–544.

Gibson, R.N. 1971. Factors affecting the rhythmic activity of *Blennius pholois*. *Anim. Behav.*, 19: 336–343.

Gibson, R.N. 1982. The effect of hydrostatic pressure cycles on the activity of young plaice *Pleuronectes platessa*. *J. Mar. Biol. Ass. UK*, 62: 621–635.

Graham, J.M., Bowers, R. and Gibson, R.N. 1987. A versatile tide machine and associated activity recorder. *J. Mar. Biol. Ass. UK*, 67: 709–716.

Green, J.M. 1971a. Field and laboratory activity patterns of the tidepool cottid, *Oligocottus maculosus*. *Can. J. Zool.*, 49: 255–265.

Green, J.M. 1971b. High tide movements and homing behavior of the tidepool sculpin *Oligocottus maculosus*. *J. Fish. Res. Bd. Can.*, 28: 383–389.

Harris, G.J. and Morgan, E. 1984a. Entrainment of the circatidal rhythm of the estuarine amphipod *Corophium volutator* to non-tidal cycles of inundation and exposure in the laboratory. *J. Exp. Mar. Biol. Ecol.*, 80: 235–245.

Harris, G.J. and Morgan, E. 1984b. The effects of salinity changes on the endogenous circa-tidal rhythm of the amphipod *Corophium volutator*. *Mar. Behav. Physiol.*, 10: 199–217.

Hastings, M.H. 1981. The entraining effect of turbulence on the circa-tidal activity rhythm and its semi-lunar modulation in *Eurydice pulchra*. *J. Mar. Biol. Ass. UK*, 61: 151–160.

Holström, W.F. and Morgan, E. 1983a. Laboratory entrainment of the rhythmic swimming activity of *Corophium volutator* to cycles of temperature and periodic inundation. *J. Mar. Biol. Ass. UK*, 63: 861–870.

Holström, W.F. and Morgan, E. 1983b. The effects of low temperature pulses in rephasing the endogenous activity rhythm of *Corophium volutator*. *J. Mar. Biol. Ass. UK*, 63: 851–860.

Jones, D.A. and Naylor, E. 1970. The swimming rhythm of the sand beach isopod *Eurydice pulchra*. *J. Exp. Mar. Biol. Ecol.*, 4: 188–199.

Klapow, L.A. 1972. Natural and artificial rephasing of a tidal rhythm. *J. Comp. Physiol.*, 79: 233–258.

Moore-Ede, M.C., Sulzman, F.M. and Fuller, C.A. 1982. *The Clocks that Time Us*. Harvard University Press, Cambridge, Massachusetts.

Morgan, E. 1965. The activity rhythm of the amphipod *Corophium volutator* and its possible relationship to changes in hydrostatic pressure associated with the tides. *J. Anim. Ecol.*, 34: 731–746.

Morgan, E. 1991. An appraisal of tidal activity rhythms. *Chronobiol. Int.*, 8: 283–306.

Morgan, E. and Cordiner, S. 1994. Entrainment of a circa-tidal rhythm in the rock-pool blenny *Lipophrys pholis* by simulated wave action. *Anim. Behav.*, 47: 663–669.

Naylor, E. and Atkinson, R.J. 1972. Pressure and the rhythmic behaviour of inshore marine animals. In: Sleigh, M. and Alister, G. (Eds), *The Effects of Pressure on Organisms*, pp. 395–415. Academic Press, New York.

Naylor, E., Atkinson, R.J. and Williams, B.G. 1971. External factors influencing the tidal rhythm of shore crabs. *J. Interdiscipl. Cycle Res.*, 2: 173–189.

Naylor, E. and Williams, B.G. 1984. Phase-responsiveness of the circatidal locomotor activity rhythm of *Hemigrapsus edwardsi* to simulated high tide. *J. Mar. Biol. Ass. UK*, 64: 81–90.

Northcott, S.J. 1991. A comparison of circatidal rhythmicity and entrainment by hydrostatic pressure cycles in the rock goby, *Gobius paganellus* and the shanny, *Lipophrys pholis*. *J. Fish. Biol.*, 39: 25–33.

Northcott, S.J., Gibson, R.N. and Morgan, E. 1991a. On-shore entrainment of circatidal rhythmicity in *Lipophrys pholis* by natural zeitgeber and the inhibitory effect of caging. *Mar. Behav. Physiol.*, 19: 63–73.

Northcott, S.J., Gibson, R.N. and Morgan, E. 1991b. Phase responsiveness of the activity rhythm of *Lipophrys pholis* to a hydrostatic pressure pulse. *J. Exp. Mar. Biol. Ecol.*, 148: 47–57.

Northcott, S.J., Gibson, R.N. and Morgan, E. 1991c. The effect of tidal cycles of hydrostatic pressure on the activity of *Lipophrys pholis*. *J. Exp. Mar. Biol. Ecol.*, 148: 35–45.

Palmer, J.D. 1974. *Biological Clocks in Marine Organisms: The Control of Physiological and Behavioral Tidal Rhythms*. John Wiley & Sons, New York.

Reid, D.G. and Naylor, E. 1989. Are there separate circatidal and circadian clocks in the shore crab *Carcinus maenas*? *Mar. Ecol. Prog. Ser.*, 52: 1–6.

Reid, D.G. and Naylor, E. 1990. Entrainment of bimodal circatidal rhythms in the shore crab *Carcinus maenas*. *J. Biol. Rhythms*, 5: 333–347.

Reid, D.G. and Naylor, E. 1993. Different free-running periods in split components of the circatidal rhythm in the shore crab *Carcinus maenas*. *Mar. Ecol. Prog. Ser.*, 102: 295–302.

Reid, D.G., Bolt, S.R., Davies, D.A. and Naylor, E. 1989. A combined tidal simulator and actograph for marine animals. *J. Exp. Mar. Biol., Ecol.*, 125: 137–143.

Taylor, A.C. and Naylor, E. 1977. Entrainment of the locomotor rhythm of *Carcinus maenas* by cycles of salinity change. *J. Mar. Biol. Ass. UK*, 57: 273–277.

Warman, C.G., Abello, P. and Naylor, E. 1991. Behavioral responses of *Carcinus mediterraneus* to changes in salinity. *Sci. Mar.*, 55: 637–643.

Williams, B.G. 1969. The rhythmic activity of *Hemigrapus edwardsi*. *J. Exp. Mar. Biol. Ecol.*, 3: 215–223.

Williams, B.G. 1995. Tidal biological clocks and their diel counterparts. In: Hartnoll, R.G. and Hawkins, S.J. (Eds), *Marine Biology — A Port Erin Perspective*, in press. Immel Publishing Co., London.

Williams, B.G. and Naylor, E. 1967. Spontaneously induced rhythm of tidal periodicity in laboratory-reared *Carcinus*. *J. Exp. Biol.*, 47: 229–234.

Williams, B.G. and Naylor, E. 1969. Synchronization of the locomotor tidal rhythm of *Carcinus J. Exp. Biol.*, 51: 715–725.

Williams, B.G., Palmer, J.D. and Hutchinson, D.N. 1993. Comparative studies of tidal rhythms. XIII. Is a clam clock similar to those of other animals? *Mar. Behav. Physiol.*, 24: 1–14.

Williams, J.A. 1985. An endogenous tidal rhythm of blood-sugar concentrations in the shore crab *Carcinus maenas*. *Comp. Biochem. Physiol.*, 81A: 627–631.

5

Persistent Fortnightly and Monthly Rhythms

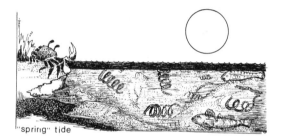

"spring" tide

I will begin this chapter with a few words about definitions. Some authors refer to rhythms that match the period of the synodic month as *lunar* rhythms. If one is concerned about accurate communication — a mandate of science — that designation is a bad choice: there are also lunar days and lunar-day rhythms. How can a reader be sure what length of rhythm is meant when just lunar rhythm is stated? I recently read a paper (the declarificationist author shall remain unnamed) that described the days of a monthly rhythm as "lunar days!" It is all rather silly coinage: no one would ever say Christmas comes in the lunar of December! Or would they look into the night sky and say, "at this time of the lunar the month-light illuminates beautifully!" There can be no confusion if we use *month* to describe these rhythms. Organismic rhythms phased to the spring or neap tides are often called semi-lunar rhythms. Again, a poor choice: the two tides/lunar day could be called semi-lunar rhythms also. Bimonthly, or the British *fortnightly* (a more romantic designation coming from a nation that once ruled the seas) removes that ambiguity. Thus, in the name of clear exposition, and contrary to most of the literature, for ease in translation I will use month and fortnight to mean month and fortnight.

By the time Korringa (1947) published his now-classic paper on the

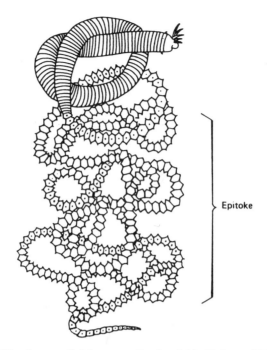

Figure 5-1   The Samoan Palolo worm *Eunice viridis* (Palmer, 1974).

subject, 34 cases of fortnightly and monthly rhythms were known. Since that time, many more have been discovered (for an informative, concise review see, Pearse, 1990). Most of the rhythms described are those of invertebrates and are associated with reproduction — events such as gametogenesis, spawning, and larval release — events that are synchronized to a particular single phase (such as full moon) of the moon if they are monthly rhythms, or to a spring or a neap tide if they are fortnightly rhythms. Both types of organismic cycles are usually "embedded" in an annual rhythm so that they are displayed at only certain seasons.

A classic example is that of the marine Samoan Palolo worm *Eunice viridis* (Fig. 5-1), which reproduces near dawn, mainly on a single day near the last quarter of the moon, in October and/or November (Hauenschild *et al.*, 1968; Caspers, 1984). In preparation for its annual orgy, the polychaete adds to its *derrière* a string of new smaller segments (collectively called an epitoke) that become packed with either green eggs or blue sperm. Then, at a given moment (so to speak), all the worms on the reef so prepared, release these foot-long string of swimming gonads that wriggle on their own up to the surface, where they perform a decerebrate dance that churns the water into a kind of frothy brothel. Up until recently, thanks to ancient lore, the Samoan natives stood ready in their canoes at just the right moment for this "gift from the sea" (modern Samoans have only to look for "Worm Day" blazoned on their calendars). When the

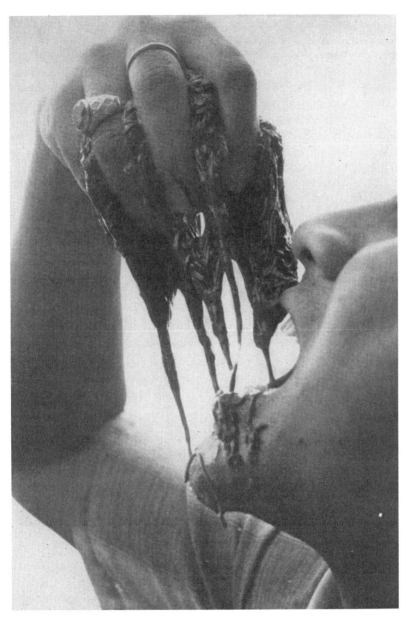

Figure 5-2 Unable to restrain himself, a vermes gourmand gauchely indulges in fresh (still wriggling) Palolo, rather than waiting until they are baked. Yes, they do momentarily stick to the roof of one's mouth (Smetzer, 1969).

lookout spots the first rising he calls out, *"Ua sau le Palolo"* (whatever that means?), and the natives rush to scoop up worm derrières by the millions. Many of these excited collectors cannot wait to get their catch home to be cooked (Fig. 5-2), but most Samoan gourmands — at roughly the time we in the United States enjoy a traditional Thanksgiving turkey — wait to dine on delicious fresh-baked Palolos, which turn spinach green in the oven. Miller and Pen (1959) give recipes and nutritional values: Palolos are low in fat, and higher in vitamin A and carotene than chicken eggs. Meanwhile, those millions of worm-ends who eluded the banquet table, finish their hootchy-kootchy and disintegrate in synchrony on the ocean surface, filling the sea with their technicolor gametes. Pronuclei fuss with abandon, and a species lacking an intromittent organ, otherwise destined to procreate only via random external fertilization, efficiently perpetuates its existence by substituting timing for orgasm. Theirs is quite a "tail." As an aside, those readers wishing to order Palolo when dining in Fiji, should look for "mbalola" on the menu.

As far as it is known, fortnightly and monthly rhythms are not particularly common in the sea. Also, a species that is rhythmic in one locale, may not be so in another. An even greater enigma is finding different populations of the same species whose rhythm is phased to different segments of the month. Why would synchronizing clues differ? Lastly, and of greatest importance as far as inclusion in this book is concerned, very few of these organismic cycles are known to persist in the laboratory, or have ever been so tested.

The Polychaete *(Typosyllis prolifera)*

As indicated by the species name, this syllid worm is a breeding machine. It undergoes a reproductive process basically the same as epitoky in the Palolo Worm, but in syllids it is called stolonization. A voyeur-like exposé of the scheduling of the worm's procreative life is presented in Fig. 5-3. Sunrise provokes the release of gamete-laden stolons that swim to the surface where the males break open releasing their sperm and a pheromone, the latter signaling the females to liberate their gametes. Eggs and sperm unite and a new generation begins. To insure intersexual communion, the kamikaze-like swarming takes place a few days before the full moon. Unlike the Palolo, this pattern repeats itself through the warmer months of the year, making it a bona fide monthly rhythm. The first 16 days of a cycle are concerned with a post-reproductive generation time (reconstitution of its *derrière*), followed by a 14-day stolon growth and release interval.

This polychaete is rather easy to maintain in the laboratory where it continues its usual dalliance. Manipulating the laboratory environment has shown that the annual rhythm is a photoperiodic, rather than annual-clock-driven event; and that the daily timing of swarming and gamete release is exogenously, rather than clock, controlled (however, that photoperiodism is involved certainly suggests that the worm may also be armed with a

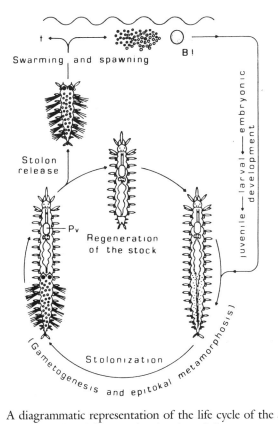

Figure 5-3 A diagrammatic representation of the life cycle of the syllid polychaete *Typosyllis prolifera*. After a stolon is released the worm grows a new posterior end (a process that takes about 16 days). Then the new derrière undergoes a 14-day metamorphosis into a stolon, a trailer-like being brimming with either eggs or sperm; note the Pv = proventriculus, which will be discussed in the last chapter. The stolon is released at dawn a few days before full moon, swims to the sea surface, and breaks open releasing its gametes (the spawn). B.1 is an egg (Franke, 1986).

solar-day clock, as has been shown for so many other animals (Saunders, 1976). The monthly rhythm is clock based.

Maintenance in the laboratory under a summer day/night cycle of 16 h of light alternating with 8 h of darkness, the worms continue to reproduce (they are long-day animals), but out of synchrony with one another. However, 2–4 nights (but not less) of simulated moonlight (a weak, 0.4 lux light is left on all night) molds the population's sex life into a rhythmic manifestation with an average period of 31 days (range for individuals = 29–34 days) (Fig. 5-4). That period is temperature compensated, i.e., it stays virtually the same in cultures kept at any temperature between 15°C and 25°C. ($Q_{10} = 1.04$) (Franke, 1986).

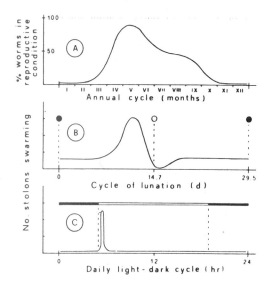

Figure 5-4 A diagram of the three rhythms of the polychaete *Typosylis prolifera:* (A) the annual one; (B) the monthly one showing stolon release a few days before the time of full moon; and (C) the daily one peaking at sunrise (Franke, 1986).

The Flatworm (Convoluta roscofensis)

This small, green, acoelous turbellarian lives buried in the intertidal sediments on the west coast of France and on the Channel Islands. It is green because harbored within its mesenchyme is a symbiotic alga upon which the worm depends for survival (*Convoluta* digests its tenants). The hungry worm must therefore move out of the sediments occasionally in order for the alga to photosynthesize; the surfacing takes place during each daytime low tide. The population density of these worms is great, so when they are up on the surface of, say, the beach by the Roscoff Marine Lab, the white sands are streaked verdant. The worms are also conspicuous because of their stench — they smell like decaying fish. When the tide returns they re-burrow (for review see Palmer, 1976).

When sediment containing *Convoluta* is brought into the laboratory and placed in constant conditions, the vertical-migration rhythm persists for a few days (Keeble, 1910).

There also appears to be a fortnightly rhythm in population size: the numbers on the surface, evaluated by how green the sediment appears, are greatest just before the spring tides, and then rapidly diminish. The cause is a halving of the size of the population, but not in a way one might expect. The posterior ends of the worms have become filled with zygotes, and at this time each worm jettisons its *derrière* and abandons it beneath the sand where young develop, while the pared anterior remnant of the worm

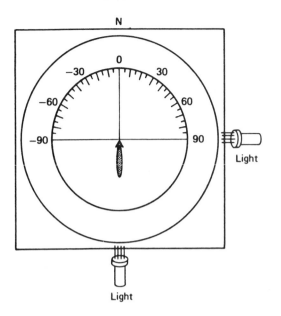

Figure 5-5 The orientation testing platform used in measuring the turning behavior of the flatworm *Dugesia dorotocephala*. The surface is illuminated from two directions, so that these worms, who always swim away from light, are forced into the left quadrant of the grid. The sector they are in as they cross the arc line is recorded for each worm (Brown *et al.*, 1970).

continues its commute to and from the surface to enable its symbionts to sunbathe. The persistence of this fortnightly reproductive rhythm has yet to be tested in the laboratory (Keeble, 1910).

Another Flatworm *(Dugesia dorotocephala)*

By pure accident this platyhelminth was found to possess a monthly clock. The experiment that revealed the timepiece was designed to learn if this little planarian could respond to the earth's magnetic field, and to an augmented one. The worm is photonegative and in its attempt to escape from two point sources of light, was forced to swim over a polar grid (Fig. 5-5). The path each worm chose was quantified by recording the number of the sector it was in as it crossed the arc periphery. After a few control animals ran this gauntlet, a magnetic field was imposed, the same animals run again, and any change in orientation noted. The measurements were made daily for several years. A basic problem was encountered: the day-to-day and week-to-week variability was surprisingly large.

To lessen the magnitude of this difficulty the data were reduced to the daily difference between controls and experimentals. The end result was one of the first demonstrations that an organism was sensitive to geomagnetism (Brown, 1962) and could use the field for orientation (the same

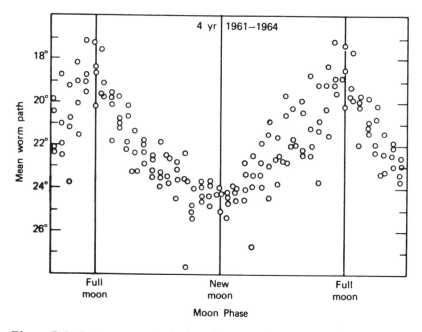

Figure 5-6 A 4-year record of orientation (away from two point light sources) of the freshwater planarian *Dugesia dorotocephala*. Each point is the mean path taken by 45 worms in the apparatus described in Fig. 5-5. Other than the times when the worms' orientations were being measured they were kept in constant darkness and at constant temperature (Brown, 1969b).

apparatus and worm were used to demonstrate a sensitivity to electrostatic fields (Brown, 1969b) and gamma radiation (Brown & Park, 1964)). While milking the data for all it contained, the daily mean paths of the control animals happened to be plotted and a very clear monthly rhythm in orientation was discovered (Fig. 5-6). During the days centered over new moon, the worms deviated significantly counterclockwise to the route they took at full moon (Brown, 1969a; Brown & Park, 1975). Other than the short time that the planarians were observed in the orientation test chamber, they were kept in constant darkness — meaning that they received no obvious clues to the interval of a month. It is hard to even imagine what adaptive significance there would be for this organism to have a monthly clock: it is a freshwater species, that inhabits small, atidal, pools and ponds. The monthly rhythm in *Dugesia* has been confirmed by Goodenough (1978), however, maximum counterclockwise turning occurred near full moon, rather than at new moon. The reason for the difference is unknown, but the experiments did differ in that Brown and Park made all their measurements in the morning, while Goodenough, somewhat of a slugabed, made her's in the afternoon.

In two other studies of the role of magnetic orientation, monthly rhythms were serendipitously discovered. To quantify pigeons' homing ability, investigators would carry birds from their home lofts to a good distance away and release them. As the birds flew away the investigators watched them until they disappeared and then noted each bird's vanishing bearing. When all the birds in a sample had been released, a "mean vanishing bearing" (MVB) was calculated. In repeating this procedure, when the same birds were released from the same site, it was found that the MVB differed slightly from day-to-day (but the birds still got home), and that this variation correlated with the natural fluctuations of the earth's magnetic field (Keeton *et al.*, 1974). Further careful examination of the data revealed that the MVB varied in an orderly fashion with a period of 30 days (Larkin & Keeton, 1978).

In another study, this one using the nudibranch *Tritonia diomedea*, when a sample population remained undisturbed in an arena, the mollusks, like the Sphinx, faced east. When the earth's magnetic field in the arena was cancelled, the nudibranch sample assumed a random orientation. In a 4-month study consisting of near daily observations, with the mollusks in the natural geomagnetic field, it was found that they did not always face east. A monthly rhythm in orientation was discovered, with the mollusks facing northeast around the times of full moon and north-northwest around the third quarters of the moon (Lohman & Willows, 1987).

The Land Crab (*Sesarma haematocheir*)

This terrestrial crab, rather common in Japan, must deposit its larvae in the sea. It lives in many different habitats, and each separate population appears to have its own means of adjusting its reproductive life to the dictates of its particular locale. I will discuss only a few of these sub-populations, which together make a truly unusual story.

The narrative begins with groups of crabs living on a hillside above the Ogamo River (near Kyoto). Mother Nature has endowed to this landlubber a real challenge in completing its life cycle. In addition to living on a hillside, the river into which it places larvae is freshwater (which is deadly to the young), and the trip down river to the sea is over 100 m. This *Sesarma* copulates during the summer months. The zygotes are extruded and stick to hairs on the ventral side of the female's abdomen. When development reaches the zoea stage, the females trudge down to the river's edge in the late afternoon, and at dusk wade out into the water a short way, get a good grip on stones, and vigorously wave their abdomens up and down in a whale swimming motion (Fig. 5-7). This causes the zoeae to smash through the egg membrane and start their dangerous ride downstream to the sea and the resuscitating salt water. A female's form of the miracle of birth completed, she wades back to *terra firma* and deftly uses her dangerously powerful chelae to carefully clean the "afterbirth" from her abdomen, in preparation for exuding a new batch of zygotes (Saigusa & Hidaka, 1978).

The point here, however, is the timing. Larval release occurs only at

Figure 5-7 Zoeae release behavior of the terrestrial crab, *Sesarma haematocheir* (Saigusa, 1982).

dusk, and only on the days around the new and full moons. What are the time cues? Dusk is the light off . . . that is easy. If the release took place only at full moon, that would be easy too: the presence of the moon would be the suspected signal. But this is a fortnightly occurrence. What governs it?

Ovigerous females were brought into the laboratory and maintained under a constant photoperiod of 14-h of light/10-h darkness and 23 ± 1°C. The fortnightly release rhythm persisted for six cycles with a circa-period of something short of 15 days. Larvae release appears to be clock controlled.

In the next experiment, the same photoperiod was retained, but now "moonlight" was added. The intensity of the artificial moonlight was 2–4 lux, but it was not offered by simply substituting this intensity for the otherwise 10-h scotophase for 5–6 nights centered on the time of full moon in the heavens outside of the laboratory. Instead, using moonrise/moonset tables, the light was turned on and off at those times, meaning that the ersatz moon "rose" on average, about 50 min later each night over the lab-crabs. This complex illumination offering is portrayed in Fig. 5-8, where it is also seen that the phase of full moon was delayed by 7 days from the natural one. When subjected to this combination, the crabs displayed

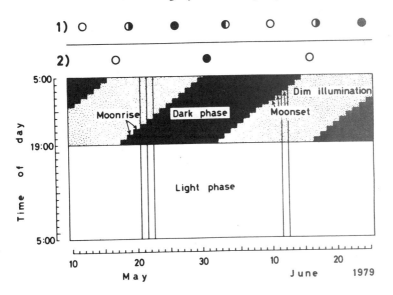

Figure 5-8　The complex photoperiod offered to *Sesarma haematocheir* to reset the phase of its fortnightly larval-release rhythm. The fundamental photoperiod is light on from 5 a.m. to 7 p.m., and darkness thereafter. But on certain days the 7 p.m. to 5 a.m. scotophase is interrupted by 2–4 lux illumination that is a surrogate for the full moon. Once begun, each night "moonrise" is scheduled about 50 min later. (1) across the top of the figure gives the phase changes of the moon in nature. (2) shows that the moon-phase cycle has been delayed in the experiment by 7 days (Saigusa, 1980).

an exact 15-day release rhythm phased 7 days after the natural cycle outside of the laboratory (Saigusa, 1980). The conclusions of these two studies is that the fortnightly rhythm is either controlled by a clock or that the crabs can count off 15-day intervals; and that the rhythm can be entrained by artificial moonlight cycles.

　　Now to a field study of a different population that also lives on land, but right next to the sea. This one also shows a fortnightly larva-release rhythm, but instead of the release taking place only at dusk, it is scheduled to occur at night-time high tides around the times of the syzygies (Fig. 5-9) (Saigusa, 1982). The phasing of this rhythm in *Sesarma* is thus like the curious, vertical-migration rhythm of *Hantzschia* (Chapter 3), except that it is phased to the night-time tides while those of the alga occur only in the daytime.

　　The rhythms were tested in constant darkness and $21 \pm 1°C$. As seen in Fig. 5-10, in crabs collected on 7 July 1983, the first larval release took place around the expected times of new moon and *night-time* high tides (the latter is very interesting; other studies of terrestrial crabs have found only persistent *diurnal* solar-day based rhythms to be present (Bliss & Sprague, 1958; Palmer, 1971)). Note that the rhythm is a lunar-day one.

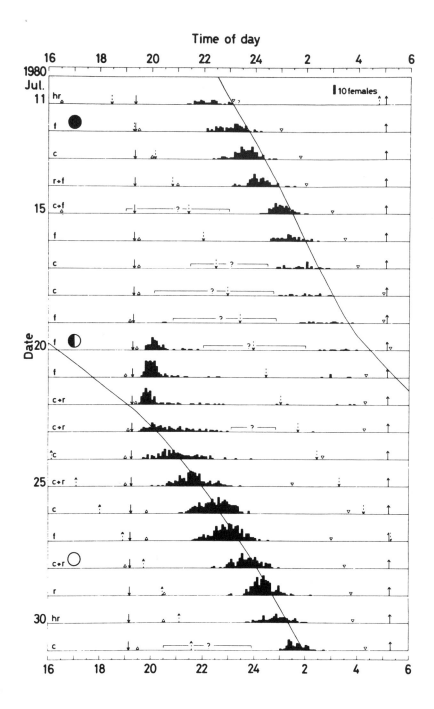

Time of day

Date

146

The second larval release appeared on schedule at expected full moon and at night, but the previous apparent association with the outside tidal exchange had broken down (Saigusa, 1986). The latter would be expected, now that we know that different individual subjects usually describe different circa lengths for their tide-associated rhythms.

Next, constant darkness was replaced with light/dark cycles that were offered at a variety of different times other than the natural one. One typical result is shown in Fig. 5-11: When "night" was scheduled to occur between 0100 h and 1100 h, larval release, tidal pattern included, moved to that time interval also! Because tide-associated rhythms are not entrained by light/dark cycles, it must have been a solar-day clock that was rephased, and that clock appears to be strongly coupled to the lunidian clock, which is simply dragged along. Curious is it not?

The conundrum here is what would set the phase of the lunar-day clock to the local tides. The crabs only venture down to the intertidal to release their larvae, and spend the rest of their time on dry land. In the chance that surf sounds and vibrations may be felt as far inland as this population's habitat, Saigusa (1986) played 12.4-h recordings of surf sounds and offered tidal turbulence cycles to lab-crabs, but found both were without entraining influence.

Another problem is the fact that artificial moonlight cycles offered only once a month, will entrain a twice/month display. One possibility, suggested and rejected by Saigusa, is that the larval-release rhythm is actually a monthly cycle, but that a single *Sesarma* population is made up of two kinds of individuals: one that releases during new moon, and the other that releases only at full moon. A casual observer counting releasing crabs on the shoreline, who does not know of this difference in individuals, would undoubtedly conclude that the rhythm is a fortnightly one. But as will be seen with the rhythms of other organisms, a single stimulus will initiate a fortnightly rhythm, and set its phase; after which the rhythm will continue to run — but usually at a circa frequency. Saigusa (1986) has just repeated the same stimulus every other cycle, and in doing so keeps the fortnightly rhythm at its basic 15-day period.

Figure 5-9 The results of a field study of the fortnightly, solar-daily, and lunar-daily rhythms of zoeae release in a population of *Sesarma haematocheir* crabs that live in the supra-tidal zone. The histograms represent the numbers of female releasing larvae. The rising and falling solid arrows represent sunrise and sunset respectively. The up and down open triangles indicate when observations began and ended each day. The diagonal lines connect the times of the high tides. Question marks are times of thinking. Moonrise and moonset are indicated by the up and down broken arrows. The large solid and open circles on the left are new and full moon. Ignore the other symbols (Saigusa, 1982).

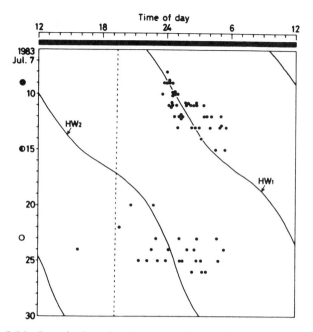

Figure 5-10 Larval release by the crab *S. haematocheir* in constant darkness and 21° ± 1°C, in the atidal laboratory. Each point is one female releasing larvae; the other symbolism is the same as in the previous figure. The releases around the days of expected new moon show the strong association with one, expected high-tide phase. By the time of the next fortnightly release, during the days around expected full moon, the different length circalunidian rhythms of individuals caused the scatter around the expected high tides (Saigusa, 1986).

The Land Crab *(Cardisoma guanhumi)*

This terrestrial decapod often lives several kilometers away from the sea, but must release its young into the ocean. During the June to early December reproductive season, a few days before full moon, often huge masses of ovigerous females begin the sometimes long (5–7 km) march to the seaside. Sometimes there are so many migrants that they stop some traffic when crossing a road (the non-animal-lover drivers just grind them into the pavement). Many make the trip in just one night, and if it takes longer than two nights the egg masses on their abdomens disintegrate. It is during the three nights centered on full moon that the crabs deposit their larvae in the sea, and then return to their inland habitats (Gifford, 1962). Larval release has not been tested in constant conditions, so it is not known if a clock is involved. It is known, however, that the species has a fine circadian clock (Palmer, 1971).

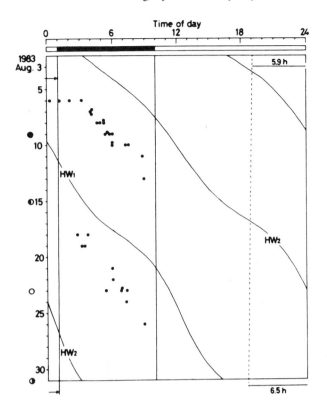

Figure 5-11 The rephasing of the lunar-day larval release rhythm in *S. haematocheir* by a phase delay in the light/dark cycle. Symbols as in Fig. 5-10. See text for description (Saigusa, 1986).

The Fiddler Crab (Uca)

In the summer months, 5–7 days before the spring tides, female *U. pugilator* become sexually receptive, and thus willing to follow a male into his burrow to begin a 2-day subterranean honeymoon. About 14 days later, at night, gravid females move into the water's edge and wave their abdomens, releasing up to 90,000 zoeae in an interval of approximately 10 min (Christy, 1978). Wheeler (1978) reports that *U. pugnax* releases larvae at new and full moons; the zoeae completed their development a synodic month later.

Ovigerous *U. pugnax*, *U. pugilator*, and *U. minax* females brought into a tide-free laboratory and maintained under day/night conditions and 25°C, released their larvae 1–2 h after expected night-time high tides, in some cases after 17 days in the laboratory. In the laboratory they also release only at the times of new and full moon (DeCoursey, 1979, 1983; Bergin, 1981).

The Amphipod *(Talitrus saltator)*

This supra-littoral amphipod has been shown to express a *circadian*, nocturnal activity rhythm. The rhythm will persist in constant darkness for as long as 46 days, and throughout this interval, the form of its cycle varies in a regular manner with maximum daily activity occurring 5–7 days after new or full moons, in synchrony with the falling spring tides. Because this display is present in constant conditions, it is thought to be controlled by a fortnightly clock. There is no sign of an accompanying tidal frequency in constant conditions, nor can one be instilled in the laboratory by subjecting animals to 7 simulated "tides," created by imposing cycles consisting of 2 h of "vibration" given at 12-h intervals (Williams, 1979).

The Isopod *(Eurydice pulchra)*

The tidal swimming rhythm of this animal was described in Chapter 3. Subsequent to those studies, improvements in laboratory techniques have made longer-term investigations possible. In one 56-day examination of animals maintained in constant darkness and 15°C, it was found that during the days of the first expected spring tides the organism's prominent cycle had a period of about 12 h. During the time of the next expected spring tides, the swimming pattern had become a mixture of two significant periods, one of 12 h, and the other 25 h. During the last subjective springs only a 25.5-h period was displayed. But the amount of daily (i.e., per 24-h intervals) activity described a very distinct fortnightly rhythm (Fig. 5-12). Periodogram analysis indicated a significant period in the 14–15 day range (Reid & Naylor, 1985).

The fortnightly rhythm persists in constant conditions, and, hence, it is thought to be under the control of its own clock. That being the case, the next question requiring an answer was what sets the phase of this rhythm in nature? At the North Wales collecting site, spring high tides are phased to midday and near midnight. A hypothesis was conceived that the timing of the tides relative to the solar day was the phase setter for the fortnightly rhythm. The test of this supposition went as follows. Freshly collected animals were exposed to eight cycles of 2 h of swaying on a flask shaker alternating with 10 h of non-agitation. In separate experiments run in a day-night cycle, the intervals of shaking were begun at 2 a.m. and 2 p.m., 4 a.m./4 p.m., 6 a.m./6 p.m., 8 a.m./8 p.m., 10 a.m./10 p.m., and 12 a.m./12 p.m. After 4 days of each of these treatments, the animals were released into constant conditions and the temporal aspects of their swimming activity measured. The maximum daily activity was produced by the 2 a.m./2 p.m. timing — just what the hypothesis predicted.

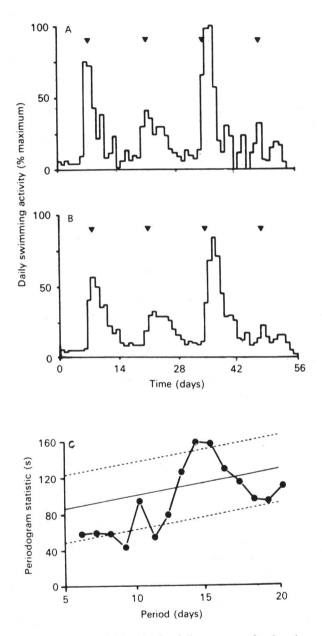

Figure 5-12 A fortnightly rhythm in the daily amount of swimming activity in *Eurydice pulchra* maintained in constant darkness and 15°C, for 56 days. (A) raw data (the solid triangles are the times of the expected highest spring tides); (B) same data after a low-pass filtering; (C) periodogram of data in (A). The dashed line in the latter indicates the 0.05% level of significance. This is a very clear demonstration of a persistent fortnightly rhythm (Reid & Naylor, 1985).

As a further test, the whole series was repeated with *Eurydice* collected from South Wales where the spring high tides were phased to occur at *dawn* and *dusk*. Again freshly collected animals were subjected to 4 days of agitation pulses timed exactly as they were for the North Wales animals. With these isopods, it was the 6 a.m./6 p.m. shaking schedule that produced the daily maximum activity — again the results presaged by the hypothesis. These results, plus others based on variations of the same experimental design, quite convincingly describe at least one manner in which the phase of this animal's fortnightly activity rhythm is set (Hastings, 1981; Reid & Naylor, 1986).

Bees

Field observations have shown that the nocturnal bee *Sphecodogstra texana* tends to fly only at dawn and dusk until a few days after the new moon. Then it extends its activity into the night-time, becoming most active just after the night of full moon. This monthly rhythm has not been tested in constant conditions (Kerfoot, 1967).

The activity of the honey bee *Apis mellifica carnica* from Morocco has been studied in constant conditions (28°C, light from a fluorescent lamp, 50% relative humidity, and a permanent food supply) in a bee-rearing room. A photoelectric counter mounted at the hive entrance tallied the bees' comings and goings. During the winter months the hive had a monthly rhythm with peaks around the times of full moon and troughs near new moon. But in summer the rhythm shortened to a fortnightly interval with peak activity at new and full moon, and minima at the quadratures (Oehmke, 1973).

The Ant Lion *(Myrmeleon obscurus)*

This animal is neither an ant nor a lion — it is a neuropteran insect that is affectionately called a "doodlebug" by the world authority (Lionel Strange) on the group's taxonomy. The young larval stage (Fig. 5-13) does, however, eat ants when opportunity knocks. The larvae dig funnel-shaped pits in loose sand and bury themselves, head up, with their large, powerful mandibles just beneath the sand surface at the bottom of the pit. When an ant, or any other small wanderer, drops into the pit, in one swift movement the huge, hollow jaws burst up out of the soil and drag the clumsy victim toward Hades, where its life juices are sucked out.

So much for the macabre part. When the larvae are brought into the laboratory and placed onto suitable soil they dig their pits and assume their underground vigil. If an investigator fills in a pit each day, by the next day a new one has been dug. When one measures the daily pit volumes, it is found that they are much larger during the days around full moon than near the new moon (Fig. 5-13). The rhythm will persist in constant darkness for 2 months, or constant illumination for 52 days, at the end

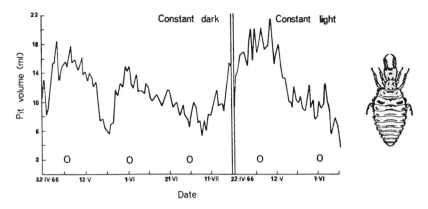

Figure 5-13 The mean-daily pit volume of 24 *Myrmeleon obscurus* larvae kept (left) in constant darkness and at 29 ± 1°C for 92 days; and another group of 24 kept (right) in constant light for 52 days. Each doodlebug was fed one ant per day. Open circles represent nights of full moon (modified from Youthed & Moran, 1969).

of which the larvae metamorphose into pupae and finally flying adults. Digging activity is also rhythmic, beginning about 4 h after expected moonrise, and only at expected night, meaning that solar-day and lunar-day clocks are also at work (Youthed & Moran, 1969). The monthly rhythm has been confirmed in my laboratory (Palmer & Goodenough, 1978).

An Intertidal Midge *(Clunio marinus)*

This tiny insect has a very complicated sex life, living most of its existence as a larva totally submerged in seawater at the lowest levels of the intertidal. Just before a spring tide arrives, mature larvae pupate, and 3–5 days later emerge as adults in the afternoon when they are finally exposed to air by a spring low, low tide. The male escapes from his puparium before the female, and after inflating his wings and completing the other tarting-up necessities required to function on dry land, he flies to a female and assists her emergence into the adult world. She is wingless so he picks her up and flies her to a spot he deems appropriately intimate and there they mate. She deposits her eggs and they both die. (If Shakespeare had known some entomology, one of his plays might have included 6-legged lovers.)

Stock cultures of this marine midge kept in the laboratory in artificial day/night cycles of 12 h of light alternating with 12 h of darkness show no predilection toward performing a fortnightly rhythm. However, simulating the light of the full moon at 2–4 consecutive nights (0.4 lux light left on during an entire dark period) will initiate the rhythm: about 22 days after the last "moon off" a peak of emergence is displayed, and it is followed by about three more at 18–22 day intervals (Fig. 5-14)

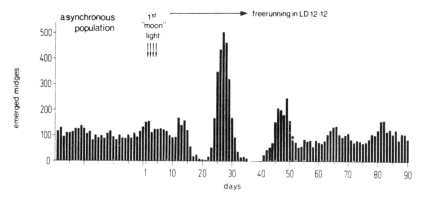

Figure 5-14   The initiation and phase setting of the fortnightly rhythm in a culture of the midge *Clunio marinus* maintained in 12-h light/12-h dark cycles. Four nights of simulated full moon (0.4 lux), given at the times of the rain of falling arrows, initiated the rhythm and set its phase (modified from Neumann, 1976).

(Neumann, 1976). Notice that the period is a circa one. In addition to a 12/12 cycle of light/ dark, a 11/11 h one will also work (Neumann & Heimbach, 1985). The display is called a "gating" response: the moonlight starts a fortnightly clock ticking and sets its hands; the clock thus running, opens a gate for pupation at a proper time if a larva is sufficiently mature. If not yet mature the gate remains closed, a larva continues on to maturity, and if it should reach it, say, in midcycle, development is arrested until the clock opens the gate again at the approach of the next spring tide.

In a related species *C. tsushimensis*, the tenacity of the period has been tested in various constant temperatures ranging between 14°C and 24°C, and found to be almost immutable ($Q_{10} = 1.06$) (Neumann, 1987).

The fortnightly rhythm just described is exhibited in light/dark cycles, thus it is possible that the midges accomplished the feat by counting the number of days that have passed. That possibility could be ruled out by testing the persistence of the rhythm in constant darkness or light. But *Cluino* feeds on algae in the culture that soon die in darkness resulting in the midges starving to death; in constant light there is no way to simulate moonlight. The solution was to expose a stock maintained under light/dark cycles to four moonlit nights, return them to light/dark for a few days, and then put them in constant illumination. As seen in Fig. 5-15, the rhythm was initiated — as expected — and then persisted in constant light (Neumann, 1976).

The above results were derived from work on a population from the northern coast of Spain (*c.*43° N latitude). It was subsequently found that artificial moonlight did not set the fortnightly rhythm of populations of *C. marinus* living above 49° N latitude: in the summer there the nights are very short and the moon never gets very high in the sky. In these northerly populations, it turns out that entrainment is produced by the

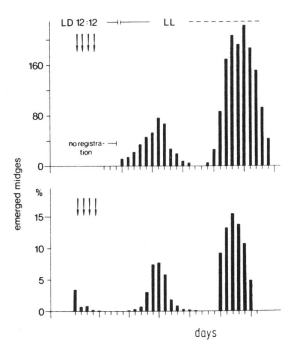

Figure 5-15 *Top:* The fortnightly eclosion rhythm of the intertidal midge *Clunio marinus*. An arrhythmic stock culture, maintained in a light/dark regimen of 12 h each and a constant temperature, was subjected to four nights of simulated moonlight (12 h of 0.4 lux artificial light; indicated by the rain of falling arrows), returned to the previous light/dark cycle for a few days, and then switched to constant light (intensity not given). The rhythm was initiated by the simulated interval of full moon, and persisted in constant conditions. *Below:* Midges kept in 12-h light/12-h dark cycles for comparison (Neumann, 1976).

timing of the tides relative to the day/night cycle. The details of this were elucidated in the laboratory using the unusual mechanical, tide-machine simulator.

Paddles attached to a motor-driven wheel, slapped the water surface as the wheel rotated. When running, the machine produced a disturbance in the water, an underwater sound (20–30 dB), and caused the bottom to vibrate (50–250 Hz). Sample cultures in a day/night cycle of 16 h of light and 8 h of darkness, were exposed to an artificial tidal cycle consisting of 8 h of agito-acousto-vibro (here abbreviated as the ooo phase) alternating with a 4.4-h interval during which the machine was turned off. As can be seen in Fig. 5-16, in a population of midges from Helgoland (54° N), each time the ooo phase corresponded to the midday of the light/dark cycle, eclosion occurred. The exact phase relationship between day and tide is

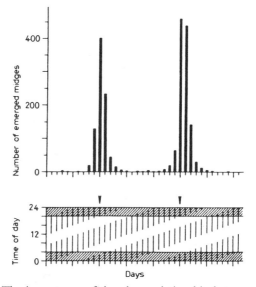

Figure 5-16 The importance of the phase relationship between an artificial tidal cycle and a light/dark cycle consisting of 16 h of light and 8 h of darkness, on the fortnightly rhythm of the midge *Clunio marinus*. The tide simulation was a complex of 8 h of agitation, noise, and vibration, alternating with 4.4 h of peace (the light/dark cycle, with the intervals of high tide indicated by the diagonal stream of bars, subtends the illustration of the rhythm). Notice that eclosion peaks were produced each time the simulated high tide occurred around midday (Neumann, 1987).

genetically determined and differs from population to population (Fig. 5-17) (Neumann, 1978).

Twelve hour and 25 minute temperature cycles, used in association with light/dark cycles (as carried out above) were found to work as well as did ooo cycles. Two methods of delivery were tried: (1) a 1.5-h spike consisting of an increase from 17°C to 21°C. in the first 15 min, and a drop back to 21°C during the next 75 min of the spike, was given every 12.4 h; or (2) a slow rise and fall by the same amount given every 12.4 h. Both means of delivery were equally effective. Fig. 5-18, shows that when temperature spikes were offered just after midday they caused maximum eclosion (Neumann & Heimback, 1984).

Epilogue

Other examples could be given, but enough have probably been presented to give one a feel for the subject. Additional information is presented in my earlier monograph (Palmer, 1974), and more current reviews can be found in: Reaka (1976), Neumann (1981, 1987), DeCoursey (1983), Naylor (1989), and Pearse (1990).

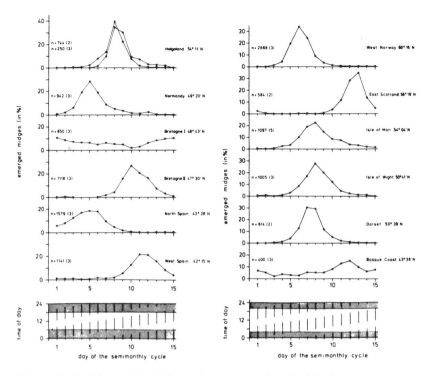

Figure 5-17   The results of the same treatment described in the Fig. 5-16 legend for different populations of the marine midge *Clunio marinus* (Neumann, 1979).

Literature Cited

Bergin, M.E. 1981. Hatching rhythms in *Uca pugilator*. *Mar. Biol.*, 63: 151–158.

Bliss, D.E. and Sprague, P.C. 1958. Diurnal locomotor activity in *Gecarcius lateralis*. *Anat. Rec.*, 132: 416–417.

Brown, F.A. 1962. Response of the planarian *Dugesia*, and the protozoan, *Paramecium*, to very weak horizontal magnetic fields. *Biol. Bull.*, 123: 264–281.

Brown, F.A. 1969a. A hypothesis for extrinsic timing of circadian rhythms. *Can. J. Bot.*, 47: 287–298.

Brown, F.A. 1969b. Response of the planarian *Dugesia*, to very weak horizontal electrostatic fields. *Biol. Bull.*, 123: 282–294.

Brown, F.A. and Park, Y.H. 1964. Seasonal variations in sign and strength of gamma-taxis in planarians. *Nature*, 202: 469– 471.

Brown, F.A. and Park, Y.H. 1975. A persistent monthly variation in responses of planarians to light, and its annual modulation. *Int. J. Chronobiol.*, 3: 57–62.

Brown, F.A., Hastings, J.W. and Palmer, J.D. 1970. *The Biological Clock: Two Views*. Academic Press, San Diego.

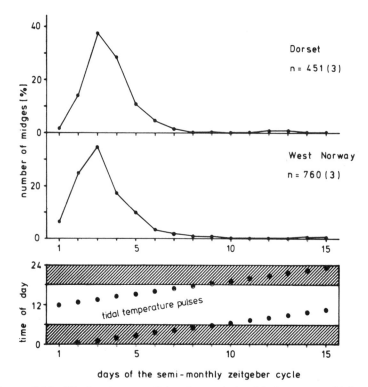

days of the semi-monthly zeitgeber cycle

Figure 5-18   The importance of the phase relationship between a tidal temperature cycle and a light/dark cycle consisting of 12 h of light alternating with 12 h of darkness, on the fortnightly rhythm of the midge *Clunio marinus*. A 90-min temperature pulse of 4°C (indicated by the heavy dots) was repeated every 12.4 h; when the cold pulse was given just after midday, it caused maximum eclosion (Neumann & Heimbach, 1984).

Caspers, H. 1984. Spawning periodicity and habitat of the palolo worm *Eunice viridis* in the Samoan Islands. *Mar. Biol.*, 79: 229–236.

Christy, J.H. 1978. Adaptive significance of reproductive cycles in the fiddler crab, *Uca pugilator*: A hypothesis. *Science*, 199: 453–455.

DeCoursey, P.J. 1979. Egg-hatching rhythms in three species of fiddler crabs. In: Naylor, E. and Hartnoll, R.G. (Eds), *Cyclic Phenomena in Marine Plants and Animals*, pp. 399–406. Pergamon, New York.

DeCoursey, P.J. 1983. Biological timing. In: Vernberg, F.J. and Vernberg, W.B. (Eds), *The Biology of Crustacea. Vol. 7*, pp. 107–162. Academic Press, San Diego.

Franke, H-D. 1986. The role of light and endogenous factors in the timing of the reproductive cycle of *Typosyllis prolifera* and some other polychaetes. *Amer. Zool.*, 26: 433–445.

Gifford, C.A. 1962. Some observations on the general biology of the land crab, *Cardisoma guanhumi* in South Florida. *Biol. Bull.*, 123: 207–223.

Goodenough, J.E. 1978. The lack of effect of deuterium oxide on the period and phase of the monthly orientation rhythm in planarians. *Int. J. Chronobiol.*, 5: 465–476.

Hastings, M.E. 1981. The entraining effect of turbulence on the circatidal activity rhythm and its semi-lunar modulation in *Eurydice pulchra. J. Mar. Biol. Ass. UK*, 61: 151–160.

Hauenschild, C., Fischer, A. and Hofmann, D. 1968. Untersuchungen am pazifishen Palolowurm *Eunice viridis* in Samoa. *Helgoländer wiss. Meeresunters*, 18: 254–295.

Keeble, F. 1910. *Plant Animals*. Cambridge University Press, Cambridge.

Keeton, W.T., Larkin, T.S. and Windsor, D.M. 1974. Normal fluctuations in the earth's magnetic field influence pigeon orientation. *J. Comp. Physiol.*, 95: 95–103.

Kerfoot, W.B. 1967. The lunar periodicity of *Sphecodogastra texana*, a nocturnal bee. *Anim. Behav.*, 15: 479–486.

Korringa, P. 1947. Relations between the moon and periodicity in the breeding of marine animals. *Ecol. Monogr.*, 17: 347–381.

Larkin, T. and Keeton, W.T. 1978. An apparent lunar rhythm in the day-to-day variations in initial bearings of homing pigeons. In: Schmidt-Koenig, S. and Keeton, W.T. (Eds), *Animal Migration, Navigation, and Homing*, pp. 92–106. Springer-Verlag, Berlin.

Lohman, K.J. and Willows, A.O. 1987. Lunar-modulated geomagnetic orientation by a marine mollusk. *Science*, 235: 331–334.

Miller, C.D. and Pen, F. 1959. Composition and nutritive value of Palolo (*Palolo siciliensis*). *Pacif. Sci.*, 13: 191–194.

Naylor, E. 1989. Temporal aspects of adaptation in the behavioural physiology of marine animals. In: Klekowski, R.Z. (Ed.), *Proceedings of the 21st European Marine Biology Symposium*, pp. 123–135. Polish Academy of Sciences, Gdansk.

Neumann, D. 1976. Entrainment of a semilunar rhythm. In: DeCoursey, P.J. (Ed.), *Biological Rhythms in the Marine Environment*, pp. 115–127. University of South Carolina Press, Columbia.

Neumann, D. 1978. Entrainment of a semilunar rhythm by simulated tidal cycles of mechanical disturbance. *J. Exp. Mar. Biol. Ecol.*, 35: 73–85.

Neumann, D. 1979. Time cues for semilunar reproduction rhythm in the European populations of *Clunio marinus*. I. The influence of tidal cycles of mechanical disturbance. In: Naylor, E. and Hartnoll, R.G. (Eds), *Cyclic Phenomena in Marine Plants and Animals*, pp. 423–433. Pergamon Press, Oxford.

Neumann, D. 1981. Tidal and lunar rhythm. In: Aschoff, J. (Ed.), *Handbook of Behavioral Neurobiology. Vol. 4*, pp. 351–380. Plenum, New York.

Neumann, D. 1987. Tidal and lunar rhythmic adaptations of reproductive activities in invertebrate species. In: Pévet, L. (Ed.), *Comparative Physiology of Environmental Adaptations*, III. pp. 152–170. Karger, Basel.

Neumann, D. and Heimbach, F. 1984. Time cues for semilunar reproduction rhythms in European populations of *Clunio marinus*. II. The influence of tidal temperature cycles. *Biol. Bull.*, 166: 509–524.

Neumann, D. and Heimbach, F. 1985. Circadian range of entrainment in the semilunar eclosion rhythm of the marine insect *Clunio marinus*. *J. Insect Physiol.*, 31: 549–557.

Oehmke, M.G. 1973. Lunar periodicity in flight activity of honey bees. *J. Interdiscipl. Cycle Res.*, 4: 319–335.

Palmer, J.D. 1971. Comparative studies of circadian locomotory rhythms in four species of terrestrial crabs. *Am Midl. Nat.*, 85: 97–107.

Palmer, J.D. 1974. *Biological Clocks in Marine Organisms.* Wiley-Interscience Publications, New York.

Palmer, J.D. 1976. Clock-controlled vertical migration rhythms in intertidal organisms. In: DeCoursey, P.J. (Ed.), *Biological Rhythms in the Marine Environment*, pp. 239–255. University of South Carolina Press, Columbia.

Palmer, J.D. and Goodenough, J.E. 1978. Mysterious monthly rhythms. *Nat. Hist.*, 87: 64–69.

Pearse, J.A. 1990. Lunar reproductive rhythms in marine invertebrates: maximizing fertilization? In: Hoshi, M. and Yamashita, O. (Eds), *Advances in Invertebrate Reproduction*, V. pp. 311–316. Elsevier, Oxford, UK.

Reaka, M.L. 1976. Lunar and tidal periodicity of molting and reproduction in stomatopod crustacea: a shellfish herd hypothesis. *Biol. Bull.*, 150: 468–490.

Reid, D.G. and Naylor, E. 1985. Free-running, endogenous semilunar rhythmicity in a marine isopod crustacean. *J. Mar. Biol. Ass. UK*, 65: 85–91.

Reid, D.G. and Naylor, E. 1986. An entrainment model for semilunar rhythmic swimming behaviour in the marine isopod *Eurydice pulchra*. *J. Exp. Mar. Biol.*, 100: 25–35.

Saigusa, M. 1980. Entrainment of a semilunar rhythm by simulated moonlight cycle in the terrestrial crab, *Sesarma haematocheir. Oecologia*, 46: 38–44.

Saigusa, M. 1982. Larval release rhythm coinciding with solar day and tidal cycles in the terrestrial crab *Sesarma* — harmony with the semilunar timing and it adaptive significance. *Biol. Bull.*, 162: 371–386.

Saigusa, M. 1986. The circa-tidal rhythm of larval release in the incubating crab *Sesarma. J. Comp. Physiol.*, 159: 21–31.

Saigusa, M. and Hidaka, T. 1978. Semilunar rhythm in the zoea-release activity of the land crabs *Sesarma. Oecologia* 37: 163–176.

Saunders, D.S. 1976. *Insect Clocks.* Pergamonl Press, Oxford.

Smetzer, B. 1969. Night of the Palolo. *Nat. Hist.*, 87: 64– 71.

Wheeler, D.E. 1978. Semilunar hatching periodicity in the mud fiddler crab, *Uca pugnax. Estuaries*, 1: 268–269.

Williams, J.A. 1979. A semi-lunar rhythm of locomotor activity and moult synchrony in the sand beach amphipod *Talitrus saltator*. In: Naylor, E. and Hartnoll, R.G. (Eds). *Cyclic Phenomena in Marine Plants and Animals*, pp. 407–414. Pergamon Press, Oxford.

Youthed, G.J. and Moran, V.C. 1969. The lunar-day activity rhythm of myrmeleontid larvae. *J. Insect Physiol.*, 15: 1259–1271.

6

The Elucidation
of the Clock

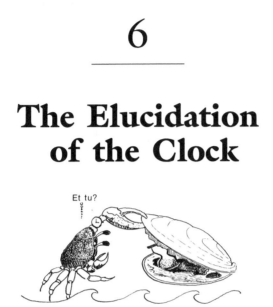

Et tu?

I will begin by highlighting the known properties of tide-associated organismic rhythms. As these rhythms persist in non-tidal conditions in the laboratory, this indicates that within an intertidal animal is a pacemaker — here called a clock — that generates the useful waveform patterns in overt behavior and physiological processes. The rhythms are innate, i.e., the period of the driving clock need not be first learned from exposure to tidal exchanges. The single cell level of organization is adequate for the expression of a tidal rhythm. When a rhythm is expressed in an atidal setting the period tends to deviate from the average period of the natural tides — a property described by the Latin *circa*, meaning about, that is used as a prefix creating words like circatidal, circalunidian and some even funnier sounding combinations. The *circa* periods are only displayed when a rhythm in unentrained. The period is temperature compensated, meaning that its length remains relatively invariable throughout a range of different *constant* temperatures. This does not mean that temperature pulses, steps, or cycles do not rephase or entrain a rhythm — they do. Other major entraining and phase setting stimuli are: hydrostatic pressure, salinity, and mechanical agitation (there are also a few esoteric, species-specific forces). Additionally, just exposing an animal that has become arrhythmic to a single tide will sometimes restart its rhythm or rhythms. Light/dark cycles, the most effective and powerful entraining agent of solar-day rhythms do not have that influence on tidal rhythms.

Ideally, the remainder of the book should describe the underlying mechanisms for all of the above. However, you are about to be disappointed: it cannot — a great deal yet remains to be learned. Maybe my next 20-year sequel, or a later one, will finally have all the answers (do not laugh, Noah, also a marine biologist, lived until he was 800 years old). What I can do for now is brief you on the state of this byway of biology in 1995.

Attacking the Clock and Coupler with Chemicals

The clock must have some bio-physico-chemical basis (because the statement includes every possibility except the supernatural, the " must" is appropriate). Following this reasoning, one attempt at deciphering the clockworks would be to test organismic rhythms with a variety of chemicals in an attempt to alter their periods (it is assumed that the overt period of a rhythm reflects the rate at which the underlying clock is running, but of course this does not have to be the case), or the phase (the hands of the clock). A broad category of substances has been tried, and we will first skim over the list of those that have been tried on circadian rhythms. Mostly, these substances have been tested using cultures of unicells because of the ease of the procedure: the compounds are simply added, one at a time, to the culture medium for a desired length of time, and later the cells are centrifuged and resuspended in fresh medium when a pulse or a sustained perturbation is to be discontinued.

Metabolic and respiratory poisons such as arsenate, cyanide, azide, p-chloromercuribenzoate, and 2,4,dinitrophenol; inhibitors of photosynthesis like DCMU and CMU; mitotic inhibitors such as urethane and fluorodeoxyuridine; growth factors such as gibberellin and kinetin; and many, many others — the list is long — were all found to have no affect on the circadian clock (Edmunds, 1988). With results like those a time had arrived to become philosophical: A clock — any clock, man-made or living — to be any good, had to have wide immunity to chemical insults from its surrounding environment if it was expected to run accurately.

Success finally began when inhibitors of macromolecular synthesis were tested. As so often happens in studies of rhythms, the early results were not clear-cut and sometimes contrary. For example, chloramphenicol, which specifically inhibits protein synthesis on 70S ribosomes, increased the amplitude (8–10 fold) of the rhythm in bioluminescence in *Gonyaulax*, but decreased the amplitude of the alga's rhythm in photosynthetic capacity (fascinating, but unimportant either way, because changes in amplitude give no information about the clockworks, e.g., a change in amplitude of a grandfather clock would mean it would only bong louder, but would still sound every hour on the hour). But compounds were eventually found that slowed the clock, e.g., cycloheximide (inhibits protein synthesis on 80S ribosomes) lengthened the period of the *Euglena* phototactic rhythm (as first discovered by Feldman, 1967). To go into greater detail is beyond

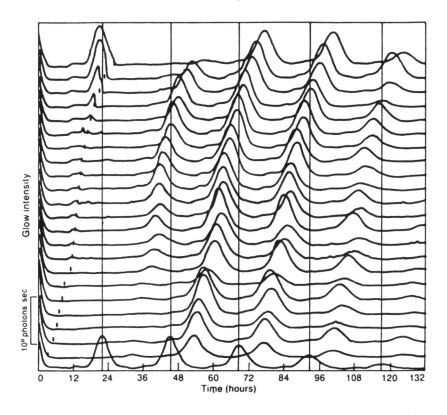

Figure 6-1 The role of 1-h 0.5μM anisomycin pulses given at various phases of the *Gonyaulax* bioluminescent-glow rhythm in constant conditions. Note that phase advances, phase delays, or no phase change were produced depending on the phase at which the pulse was given (modified from Taylor *et al.*, 1982).

the scope of this monograph (for those interested, see Chapter 4 in Edmunds's 1988 pundatorial magnum opus for a fine account) other than to say several other inhibitors are now known to produce the desired effects — period and phase changes — and that a generalization is emerging: Inhibitors of protein synthesis on 80S ribosomes are the ones that alter biological rhythms. But before we leave this category I will give one example, mainly because the data are so nice and clear, and thus the opposite of those to which we who work with tidal rhythms are accustomed. Figure 6-1, shows the results of 1-h staggered pulses of anisomycin on the *Gonyaulax* bioluminescence rhythm. (Anisomycin binds to the 80S ribosome subunit at the peptidyltransferase complex and noncompetitively blocks peptide-bond formation.) As you can see, phase advances, phase delays, or no phase changes are produced as a function of at what point in a rhythm the antibiotic is pulsed (Taylor *et al.*, 1982).

Several membrane-active agents have also been found to alter the phase and frequency of circadian rhythms. Short-chain alcohols, the ionophore valinomycin, certain ions, DES (an inhibitor of plasma membrane AT-Pase), aminophylline, theophylline, caffeine, and deuterium oxide (which seems to influence most biological processes) all are effective.

All of the substances mentioned above — and others — that produced positive effects were tested on circadian rhythms. Only a few of them have subsequently been tried on tide-associated rhythms.

Ethanol

The first substances found that would alter the period of a circadian rhythm was ethanol: it lengthened the period of the bean (*Phaseolus multiflorus*) circadian sleep-movement rhythm. A 5% solution was simply fed to the plants via a sprinkling can (Keller, 1960). Methanol was an even better period modifier (Keller, 1960; Sweeney, 1974). When added to the culture medium of a unicell (*Gonyaulax*) it shortened the period of the alga's bioluminescence rhythm, and, when pulsed, caused phase changes (Taylor *et al.*, 1979).

As with circadian rhythms, the first substance that was found to alter the period of a tide-associated rhythm was also alcohol (probably not a surprising choice to try considering the depth of the Dionysian evening crowds by the bar at our conferences). Enright (1971b) added ethanol to the aquarium in which was kept the isopod *Excirolana chiltoni*. One percent was the highest concentration any of the little teetotalers could tolerate, and mortality rates were great at that level. Ascending concentrations at

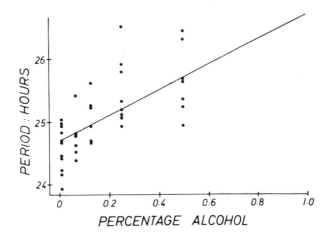

Figure 6-2 The dose-response curve for the effect of varying concentrations of ethyl alcohol on the activity rhythm of the isopod *Excirolana chiltoni* in constant conditions. The regression line has a slope of 119 min (modified from Enright, 1971b).

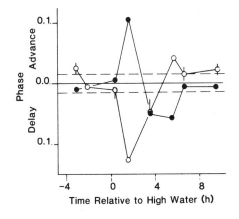

Figure 6-3 Phase-response curves indicating the influence of 3-h pulses of 0.15% ethanol (open circles), or 10^{-5} mol/l valinomycin (solid circles) on the swimming rhythm of ten-individual groups of *Corophium volutator* in constant illumination at $12 \pm 1°C$. The points are the midpoints of pulses given relative to the times of expected high water as indicated on the abscissa. Standard error bars have been added to some points. The units given on the ordinate are unstated. The broken lines are the standard deviations in phase of five untreated control groups. Alcohol pulses given during early ebb cause a delay of c.90 min; whereas pulses at low tide product 30 min advances. Roughly the opposite is true for valinomycin (Harris & Morgan, 1984a).

0.2% increments up to 0.5% were tested and a typical dose-response curve was recorded (Fig. 6-2).

Harris and Morgan (1984a) looked for similar modifying roles of ethanol on the amphipod *Corophium*. Placing groups of animals in 0.001%, 0.01%, and 0.05% ethanol, produced no changes relative to the seawater-only controls, in the periods of the persistent rhythms. However, subjecting animals to 3-h pulses of ethanol in concentrations ranging from 0.15% to 4.0%, beginning at the time of expected high tide, caused phase delays at all concentrations, the magnitudes of which increased, in general, up to a mean of approximately 175 min at a concentration of 2%, and then decreased to a mean of approximately 30 min at 4.0%. Lastly, they offered 3-h pulses of 0.15% ethanol at various phases of the activity cycle. A changing sensitivity was found ranging from a 30-min advance when offered at low tide, and a 90-min delay early in the ebb tide (Fig. 6-3) (Harris & Morgan, 1984a).

Deuterium Oxide (Heavy Water)

When heavy water (^2H_2O) became available and affordable to biologists it was first used on unicells, where it most often changed their morphology. Bruce and Pittendrigh (1960) were the first to test it on a biological rhythm, the circadian phototactic rhythm of *Euglena*. Using greater and greater concentrations over many weeks in the alga's medium they found that it lengthened the rhythm's period significantly. It was later found, using another unicell (*Gonyaulax*), that the long pretreatment was not required (McDaniel *et al.*, 1974).

Just adding heavy water to the drinking water of animals produced the same period augmenting effect (Suter & Rawson, 1968; Palmer & Dowse, 1969; Dowse & Palmer, 1972; Hayes & Palmer, 1976; White *et al.*, 1992), and would also alter the phase alignment of a rhythm to a light/dark cycle (Dowse and Palmer, 1972; McDaniel *et al.*, 1974; Hayes & Palmer, 1976).

Enright (1971a), using the tidal activity rhythm of the isopod *Excirolana*, was the first to demonstrate the period-lengthening effect induced by heavy water on a tidal rhythm. this was an easy task because the ^2H_2O could simply be added to the seawater in which the animals bathed. Heavy water lengthened the period of their swimming rhythm.

In the only other test of heavy water on a tidal rhythm, its administration was more difficult. The fiddler crabs *Uca pugnax* and *U. pugilator* were the subjects. To insure penetration, 10 μl of 99% deuterium oxide was injected into an animal through the membranous exoskeleton joint at the base of the first walking leg. Additionally, 3 ml of heavy water was spread on the floor of the participants' actographs so they had to tiptoe in it. An equal number of crabs were injected with plain distilled water, and had this added to their actograph. Each animal was first studied in constant conditions for 6 days to establish its basic rhythmic pattern, then injected with ^2H_2O, and returned to its actograph.

The amount of activity decreased in 9 of 11 experimentals by 41%; and only 4% in the controls. Period lengthening (Fig. 6-4) was caused in 73% of the experimentals, producing an average (and standard deviation) period of 26.7 ± 0.35 h; while the controls average period was 24.75 ± 0.43 h. The means differed significantly: $P < 0.001$ (Palmer, 1990a).

From these two studies on tidal rhythms (I should mention again that Enright (1976) considers the *Excirolana* clock a circadian one), and many more on circadian rhythms, it is irrefutable that deuterium oxide can alter biological rhythms. Additionally, its influence is the same on all organisms tested: it causes period lengthening and phase delays. We have no idea, however, whether it is acting directly on the clocks, on the couplers, or on something else. The problem lies in the many ways that heavy water alters biological systems. Compounds built of deuterium are more stable because deuterium forms more durable bonds than water. Its presence slows permeability. It is more viscous than water and therefore slows diffusion rates and decreases ion mobility. It alters nerve activity. The solubility of the respiratory gases is changed. And there are other effects (Thompson, 1963). This wide spectrum of influences precludes an investigator's chance of pinpointing its affect on specific parts of the

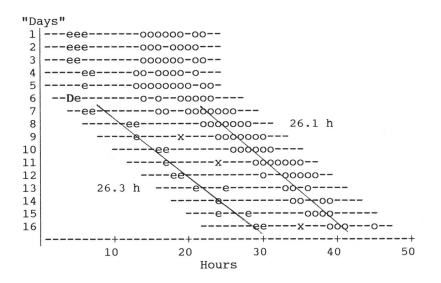

```
"Days"
 1 ---eee-------oooooo-oo--
 2 ---eee-------ooo-oooo---
 3 ---ee--------oooooo-oo--
 4 -----ee-----oo-oooo-o---
 5 -----e------ooooooo-oo-
 6 --De--------o-o--ooooo----
 7 --ee--------oo--oooooooo---
 8 -----ee--------oooooooo---  26.1 h
 9 -----e-----x----oooooooo---
10 -----ee--------oooooo----
11 -----e-------x----ooooooo--
12 ----ee----------o--ooooo--
13   26.3 h  ----e--e--------oooo-----
14           -----e--------oo-oo----
15       ----e--e-------oooo------
16         ------ee----x---oooo---o--
   --------+---------+---------+---------+---------+
         10        20        30        40        50
                        Hours
```

Figure 6-4 Compact plots of period lengthening in a single fiddler crab's (*Uca pugilator*) activity rhythm, caused by heavy water administration. On day 6, at hour marked **D**, $10\mu l$ of 99% ^2H_2O were injected into the animal and 3 ml of the substance spread on the actograph floor. The periods of both the **e** and **o** peaks increased from *c*.24 h, to the lengths indicated on the figure. The **x**s represent values that are larger than the daily hourly mean, but, because of their location, are considered to be noise (Palmer, 1990a).

clockworks — if that, in fact, is even where it is working. That is unfortunate because its influence is so uniformly predictable.

The role of deuterium oxide on the monthly orientation rhythm (Chapter 5) of the planarian *Dugesia* has been tested. The animals were maintained in 5%, 10% or 15% heavy water and their orientation to two point sources of light (Fig. 5-5, p.141) measured. The speed at which they "ran" the course was also gauged. While the deuterium caused significant retardation in the planarian's swimming rate (indicating, among other things, that heavy water had become incorporated into their bodies), it had no period-altering effect on the monthly rhythm (Goodenough, 1978). This result certainly suggests that the clock governing monthly rhythms is different from those driving circadian and circalunidian rhythms.

Azadirachtin

This triterpenoid is a garlic-smelling oil produced by the neem tree (*Azdirachta indica*), a common shade tree (belonging to the mahogany family) that lives in India. The oil is an insecticide that kills herbivorous insects directly, or stops their ability to lay eggs (Garcia & Rembold, 1984). Han and Englemann (1987) and Smietanko and Engelmann

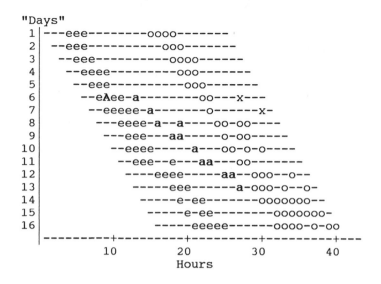

```
"Days"
  1 ---eee--------oooo-------
  2  --eee----------ooo-------
  3   --eee----------oooo------
  4    --eeee---------ooo-------
  5    --eee----------ooo-------
  6     --eAee-a--------oo---x---
  7      --eeeee-a--------o----x-
  8      ---eeee-a--a----oo-oo----
  9       ---eee---aa-----o-oo-----
 10        --eeee-----a---oo-o-o----
 11         --eee--e---aa---oo-------
 12          ----eeee-----aa--ooo--o--
 13          -----eee------a-ooo-o--o-
 14          -----e-ee-------ooooooo--
 15           -----e-ee--------ooooooo-
 16           -----eeeee------oooo-o-oo
    ---------+---------+---------+---------+---
            10        20        30        40
                     Hours
```

Figure 6-5 Compact plot showing azadirachtin-caused splitting of the activity rhythm of the fiddler crab *Uca pugnax*. On day 6, at the hour marked **A**, 10μl of azadirachtin was injected into this crab, causing the *e* peak to split off a thin branch (identified by letter **a**s). The branch subsequently fused with the *o* peak, which broadened after the impact (Palmer, 1990a).

(1989) found that injecting minute amounts of it into cockroaches (*Leucophaea*) and flies (*Musca*) caused their circadian activity rhythms to split, and to change period length — usually lengthening it.

At first thought it probably seems silly to inject an insecticide into an insect to see if it would influence its rhythms. Were it not for the results obtained with the above pests, it would seem even more foolish to do the same with crabs — even though they too are arthropods. Village idiot or not, I tried it. Azadirachtin is sold commercially as a solution of 75% ethanol, 14% neem oil, and 0.3% azadirachtin. I diluted this with 24 parts of seawater and injected 10 μl into each of a total of 30 *Uca pugnax* fiddler crabs in various phases of their activity rhythms. Ten control animals received a single 10 μl injection of a 3% ethanol solution. A variety of responses were obtained.

First, no harm was done: in Madison Avenue bumper-sticker hype, "the clocks took a licken, but kept on ticken." Within 36 hours after an injection, one of the tidal peaks split in 40% of the animals (Fig. 6-5). Period shortening was seen in 13%, and period lengthening in 7% of the crabs. Injection of two crabs caused them to become arrhythmic. Azadirachtin produced no change in 20% of the animals. The **e** peak of one of the control animals split three days before the ethanol injection, but no changes were seen after the injection in that one or any of the other controls. It must be remembered however, that the amount of alcohol administered was very small, much less than that used in studies of circadian rhythms where it was found to be a period lengthener (Palmer, 1990a).

Here would be an appropriate place to ask. "So what if azadirachtin produces these changes?" Good question; to which there is no answer; so little is known about the action of the compound it is impossible to say what its impact is on the clock and/or the coupler, or on whatever it acts to cause peak splitting. About half the time it split the **e** peak and the other time the **o** peak. While politically correct that it did not discriminate, why just one peak or the other? About all that can be said is that the results are consistent with the circalunidian clock hypothesis: only one of the two clock displays was modified.

Valinomycin

Sweeney (1974) subjected the stimulated-bioluminescent rhythm of the alga *Gonyaulax* to 4-h and 6-h pulses of this ionophore and produced small, but reproducible, phase changes in the plant's rhythm.

The tidal swimming rhythm of *Corophium* was subsequently also subjected to the ionophore. Because valinomycin is not soluble in seawater, first it must be dissolved in a non-polar solvent like ethanol, a substance already known to alter the phase of this animal's activity rhythm. Thus, in each experiment an ethanol-only control had to be run for comparison. Groups of 10 amphipods were exposed to 3-h pulses (offered at several different phases of the crustacean's rhythm) of a 10^{-5} mol/l valinomycin/0.15% ethanol mixture. Whatever change that was recorded in the alcohol controls had to be subtracted from those found with the valinomycin/ethanol combination. Figure 6-3, shows the results: a phase-response curve that is pretty much the mirror image of the one obtained with ethanol (Harris & Morgan, 1984a).

Cycloheximide

The other substance tested on *Corophium* was cycloheximide. A sustained-perturbation type of experiment was conducted: Concentrations of 0.36, 0.72, and 1.44×10^{-9} mol/l were tried. As seen in Fig. 6-6, the cycloheximide produced no important changes. At the intermediate concentration, the first two peaks were displayed, while at the highest concentration the animals became arrhythmic. In repeat experiments using the most concentrated solution, the "rhythm persisted for one or two cycles and then ceased. . . ." Note that swimming carried on undiminished in cycloheximide, suggesting that the coupler was apparently the entity sensitive to the antibiotic (Harris & Morgan, 1984a).

Enright (1971b) subjected the isopod *Excirolana* to 3 μg/ml cycloheximide, and found no effect.

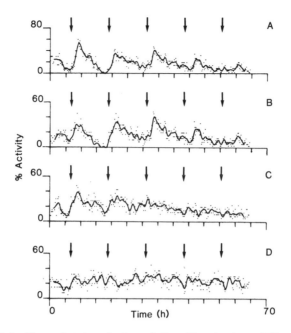

Figure 6-6   The swimming rhythm of *Corophium* in three different concentrations of the antibiotic, cyclohexomide. (A) is the control group. (B), (C), and (D) are 0.36, 0.72, and 1.44×10^{-9} mol/l solutions (respectively) of cycloheximide. Falling arrows indicate the times of expected high tides (Harris & Morgan, 1984a).

Genetic Search for the Clock in the Fruit Fly

The fruit fly, Drosophila, undergoes a very precise, persistent solar-day rhythm in which adults emerge from their pupal cases at dawn.

In 1971, Konopka and Benzer, using the mutagen ethylmethane sulfonate, induced mutants in *D. melanogaster* and found that 3 of the 2000 mutants produced had altered activity rhythms. The *period* gene (abbreviated to *per*) now comes in several varieties: *per^L*, which lengthens the rhythm to *c.*29 h in constant darkness; *per^S*, which shortens the period to *c.*19 h; and *per^O*, a null mutation that usually shows no circadian period (but often displays numerous ultradian rhythms [Dowse and Ringo, 1987]). Six other mutants have since been isolated. The wild-type, *per$^+$*, displays 24-h persistent rhythms.

The gene has been cloned and sequenced, and the *c.*1100 amino-acid sequence of its protein (called PER) deduced (Reddy *et al.*, 1983; Citri *et al.*, 1987). PER appears to be a member of a family of proteins that are transcriptional regulators, and as such is a nuclear protein in the central brain of *Drosophila* (Liu *et al.*, 1992), in a cluster of neurons known to be required for circadian rhythmicity (Ewer *et al.*, 1992).

The amount of PER exhibits a circadian rhythm in all mutants except, of course, per^O (Siwicki *et al.*, 1988; Zerr *et al.*, 1990). The amounts of *per* mRNA also oscillates circadianly (Hardin *et al.*, 1990), its phase leading that of the PER rhythm by *c.*7 h.

The *per* gene contains unusual nucleotide sequences that have also been found in spinach, rape, *Acetabularia*, chicken, mouse and human DNA (Shin *et al.*, 1985; Li-Weber *et al.*, 1987). In *Acetabularia*, the homologous sequence is in the chloroplast genome, where translation would take place on 70S ribosomes! Whether these homologues function in rhythmic processes in these other species is unknown.

Finally, per^L and the gene for *rosy-eye* color have been incorporated into a plasmid and injected into a per^O embryo in an area which will form the gonads. The resulting adults were then mated with per^O and the resulting F_1 generation flies screened for rosy eyes. These were tested and found to show long-period rhythms (Bargiello *et al.*, 1984).

No one has yet looked for the *per* gene in an intertidal organism. It should be done.

Other Clock Genes

The *frq* (= frequency) gene of the bread mold *Neurospora crassa* is a fungal sole mate of *per* (Feldman & Hoyle, 1973; Dunlap, 1993). About twenty period-length mutants are known in this ascomycete, and eight of them map to the *frq* locus.

Just recently, a mutagen was used to produce a long-period mouse. The mutant gene is called *clock*, not as unimaginative a moniker as one might first suspect. Actually, the name is an acronym standing for **c**ircadian **l**ocomotor **o**utput **c**ycles **k**aput! No, the creators did not choose the last word out of desperation so as to be able to spell timepiece. Actually it is quite appropriate: in the homozygous condition, the rhythm will persist for only about two weeks with a period of 26–29 h; but then, while the activity level is not reduced, the waveform is lost — thus, the "kaput." The clock gene maps to the midportion of mouse chromosome 5, the entire region of which is syntenic with human chromosome 4 (Vitaterna *et al.*, 1994).

Ultradian Rhythms

As described above, per^O flies do not display circadian rhythms. However, Dowse and Ringo (1987), using MESA to dissect the activity output, found that more than half of these animals had clear-cut ultradian rhythms, with periods ranging between 4 h and 18 h in length. They also found ultradian rhythms common in per^L and per^+ flies, but they are less prevalent in per^S. There was no favored ultradian period: all values were spread fairly uniformly through a 4–18 h range.

The activity rhythms of wild type, per^L and per^S have been subjected to deuterium oxide in concentrations varying between 10 and 50%. It lengthened the periods of the circadian rhythms (White *et al.*, 1992) and enhanced the expression of ultradian rhythms (Dowse & Ringo, 1992b). Curiously, the response was not dose dependent. Another interesting finding was that drinking <40% ^2H_2O extended the life span of the flies.

When strong circadian rhythms are present in flies, ultradian rhythmicity is at a minimum or missing entirely, and vice versa. This inverse relationship is one factor used by Dowse and Ringo (1987) to postulate that the reason behind our being unable to identify a circadian clock is that it does not exist. They speculate that a population of intracellular "U-oscillators" (U can stand for Uhr, or Ultradian, or Undercover, or Unproven), each with different short periods, are coupled together to produce overt circadian rhythms. They call this complex a "meta-oscillator." When they couple together strongly, a fly's circadian rhythm is robust. When a meta-oscillator is poorly coupled, a fly's activity will contain a weak circadian component, and will also include one or more "free-lance" ultradian peaks. When all U-oscillator coupling breaks down, only ultradian rhythms are displayed (as in *per*^{O}). A further postulate of these creative investigators is the suggestion that the *per*-gene product has a major role in coupling the short-period U-oscillators together (Dowse & Ringo, 1992a). We will return to this subject at the end of this chapter.

Anatomical Location of Clocks

The residences of several animal (and plant) clocks are known. The suprachiasmatic nuclei in the hypothalamus of some vertebrates; the pineal of birds; the brain in insects; the eyes of gastropods; and the eyestalks, brains and abdominal ganglia of some crustaceans all have clock-like properties. Most of the timepieces located thus far are known to drive only circadian rhythms. But that is a result of the fact that few marine laboratories have even attempted to search for tidal-rhythm clocks. A brief review of solar-day clocks will set the stage for what little is known about ocean-going clocks.

The Suprachiasmatic Nuclei (SCN)

At the base of the hypothalamus of the mammalian brain, just above the optic chiasm, lie two nuclei: the SCN. In 1972, Moore and Eichler, studying the circadian rhythm in adrenal corticosterone output; and Stephen and Zucker (1972), examining rhythmic drinking and activity rhythms — both teams using small mammals — discovered that lesioning the SCN destroyed these rhythms. As ablation experiments demonstrated, only one of the pair is required for rhythmicity to persist. Electrical stimulation of the SCN will produce phase and period changes in the locomotor rhythms of rats and hamsters (Rusak & Groos, 1982). It is important to point out, however, that not all mammalian rhythms are controlled by the SCN.

Rhythmicity destroyed by lesioning in the rate and hamsters can be restored by transplanting fetal SCN tissue into the third ventricle of the brain of these animals (Sawaki *et al.*, 1984). Even transplanting a slurry of tryspin-digested SCN cells into SCN-lesioned adults will re-establish rhythmicity (DeCoursey & Buggy, 1989; Lehman *et al.*, 1987; Silver *et al.*, 1990).

A short-period mutant (called *tau*) in the hamster produces an *c.*20-h period when present in the homozygous condition (Ralph & Menaker, 1988). When a SCN from a *tau* mutant is transplanted into a SCN-lesioned, wild-type host animal, rhythmicity is restored and the *c.*20-h period is exhibited. Interestingly, when a potential host's SCN is only partially lesioned and a *tau* mutant's SCN implanted, both the wild-type and mutant periods are displayed, sometimes within 1–2 days after the operation (Ralph *et al.*, 1990).

Before leaving this fascinating story and all its exciting implications, I feel required to comment that because of the above story, as far as the popular print media is concerned, and thus those of the public who can read, the search is over — the biological clock has been found. This of course is an egregious mistake, reporters — some of whom were undoubtedly educated at my home university — have not the slightest inkling that all plants, polychaetes, crustaceans, etc., do not have SCNs, yet do have very impressive persistent rhythms.

The Avian Pineal Gland

This romantic organ, the third eye in some lizards, and, if we are to believe Descartes, the seat of our soul, is a clock in some birds, but not in mammals. In passerine birds such as the sparrow (*Passar domesticus*), when the pineal gland is removed the circadian activity rhythm is destroyed (Gaston & Menaker, 1968), and when pineal tissue is inserted in the anterior chamber of the eye of one of these birds, the rhythmicity, describing the phase of that of the donor, is restored (Zimmerman & Menaker, 1975).

The pineal gland brings about its effect by the rhythmic release of melatonin, which is produced in the following reaction:

$$\text{Serotonin} \xrightarrow{\text{N-Acetyl-transferase}} \text{Melatonin}$$

It would be interesting to see if the pineal gland would continue its rhythmic secretion when isolated in culture, but the sparrow pineal, as one can imagine, is very small. Thus the barnyard chicken (*Gallus domesticus*) was used instead (for the same reason marine biologists like to work on lobsters, dry-land chronobiologists like to use chickens: the day's laboratory subjects can become the evening meal). It is the rate-limiting N-Acetyl-transferase activity that is followed in culture in con-

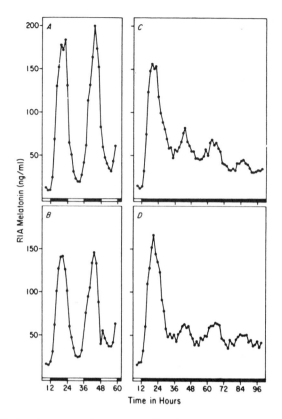

Figure 6-7   Replicate experiments showing the persistence of the
melatonin-release rhythms from two isolated chicken (*Gallus domesticus*) pineal
glands, first in alternating light/dark conditions, indicated by the empty and
filled bars at the base of (A) and (B), and then in constant darkness (filled bars)
in (C) and (D) (Takahashi *et al.*, 1980).

stant conditions. The rhythm persisted (Fig. 6-7), even in isolates as
small as one-eighth the size of the whole pineal gland (Takahashi &
Menaker, 1984) and eventually in dissociated cells in culture (Robertson
& Takahashi, 1988). But, as is so often the case, it seems that when
studying rhythms a paradox is involved. The one here is that, after learning
of the very clear rhythmic response of pineals in isolation, pinealectomy
does not affect the activity rhythm of the chicken (Takahashi & Menaker,
1979)! It turns out that the SCN is apparently the basic timer of
G. domesticus, for when it is lesioned all circadian rhythmicity is lost
(Takahashi & Menaker, 1982).

The Sea Hare (Aplysia californica)

Finally, we at last get to a marine animal. This one is a slow moving, deep
purple, gastropod that can exhibit a rather ragged circadian locomotory

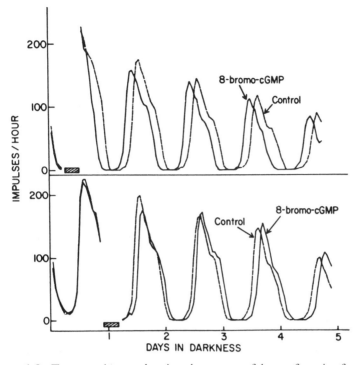

Figure 6-8 Two experiments showing the great usefulness of a pair of isolated sea hare (*Aplysia californica*) eyes, in culture, in testing the role of a substance on their spontaneous neural output. *Top:* a 6-h pulse of 8-bromo-cGMP (2×10^{-1} M) at one phase in the rhythm causes a phase advance, while, *Bottom:* the same pulse offered at another phase causes a delay (Eskin *et al.*, 1984).

rhythm (Kupfermann, 1967). A much more precise daily rhythm has been found in the spontaneous neuronal firing of the animal's eyes; first discovered by Jon Jacklet (1969). Moreover, the eye, which is only 0.5–1.0 mm in diameter, can be removed from the sea hare, along with its 1 cm-long optic nerve, and maintained alive in culture in constant conditions for as long as 2 weeks, during which time the circadian neural activity persists (Fig. 6-8). The isolate can even be entrained by a light/dark cycle (Eskin, 1971). Thus each minute eye contains a pacemaker and a mechanism for entrainment.

The eye consists of a central lens surrounded, more or less, by a retina composed of about 4000 photoreceptors, and 1000 secondary neurons. In an attempt to locate just where in the eye the clock was located, Jacklet and Geronimo (1971) surgically shaved away more and more of the eye. Eighty percent of the lens and cells could be removed without destroying the rhythmicity, however, the amplitude was progressively reduced as more of the eye was systematically cropped. Further mutilation abolished the circadian rhythmicity, but shorter and shorter ultradian rhythm appeared.

There are two eyes, each with a clock. The two are coupled together via the brain, so a phase change in one is instantly passed to the other. When a pair of eyes are isolated from one another *in vitro* and in constant conditions, they both continue, for some while, to produce identical rhythms; thus one can provide the perfect control for an experimental modification to the other (Fig. 6-8).

The above certainly demonstrates that a restricted part of the nervous system can function as an autonomous clock. Does that clock also play a role in molding the locomotor activity into a circadian waveform? The answer is a definite no and maybe. In a study of 20 *Aplysia* in constant conditions, 9 had very nice rhythms, 8 fairly good, and 2 poor ones. Out of 33 eyeless animals only 2 had very nice rhythms, 11 fairly good, and 20 were poor (Lickey *et al.*, 1977). The comparison certainly demonstrates that the eyes are not required, but when they are present, locomotor rhythmicity is improved.

The Cloudy Bubble Snail (Bulla gouldiana)

This marine mollusk is a near relative of *Aplysia*. Its clocks also reside in the optic lobes, and attempts to pin-point their locale by shaving away eye tissue has been even more informational. Below the photoreceptor layer of the eye is a layer of approximately 100 neurons — collectively called the basal retinal neurons — that are connected electrically to one another and have axons that pass into the optic nerves. Synchronous impulses from these cells constitute large compound action potentials that exhibit a circadian rhythm. Surgical removal of the overlying photoreceptor layer does not destroy the rhythm or block the clock from being phase shifted by light signals. Because all that are left after the excision are the basal retinal neurons, they must function together as a clock. In fact, when all but six of them have been removed the rhythm persists (Block & McMahon, 1984). Eventually it became possible to culture, for at least two days, single neurons and each has been found to contain its own clock (Michel *et al.*, 1993; Block *et al.*, 1993).

Two Types of Insect Clocks: Endocrine and Neural

The silkmoth *Hyalophora cecropia* undergoes an eclosion rhythm: the adults emerged from the pupal case just after dawn (or 1 h after "light on" in the laboratory). If a pupa's brain is removed, post-dawn emergence is destroyed — the acerebral adults emerge at all times of the day and night. But, when a brain is re-implanted in the abdomen of a brainless pupa, the morning eclosion phase is reinstated (Truman, 1972b). The implication is that the clock is in the brain and that it governs emergence via an eclosion hormone.

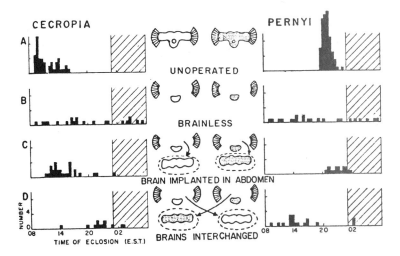

Figure 6-9 The eclosion rhythms of *Hyalophora cecropia* (left) and *Antherea pernyi* (right) silkmoths in light/dark cycles consisting of 17 h of light alternating with 7 h of darkness. (A) normal rhythm; (B) brain removal caused loss of rhythmicity; (C) brains removed and replaced in abdomen of the same animal: the rhythms persisted; and (D) brains removed and transplanted into the other species: the phase of the donor brain was carried to the recipient (Truman & Riddleford, 1970; Truman, 1972a).

The hypothesis was proven with the help of another silkmoth, *Antheraea pernyi*. This species also displays an eclosion rhythm, but emergence is phased to the last few hours before "light off." That moth was subjected to the same brain-out, brain-in manipulations as was *H. cecropia* with the same result — except for the phase difference (Fig. 6-9). Then brains from *A. pernyi* pupae were then transplanted into brainless *H. cecropia*, and vice versa, and emergence observed in a light/dark cycle. The phase of the emergence rhythm was that of the transplanted donor brains. A clock is certainly in the brain (Truman & Riddleford, 1970).

In an attempt to determine what part of the brain contained the clock, parts of it were surgically removed before transplantation. This way it was found that the lateral portions of the cerebral lobes probably contained the eclosion clock (Truman, 1973).

These moths display additionally, a circadian flight rhythm that is also controlled by a brain-based clock. It is not, however, the same clock that dictates eclosion times: While removal of the brain annihilates the flight rhythm, replacing it in the abdomen does not rectify the cyclic demise. Hormones are not involved, neural connections must be intact for this clock to rule. By surgical slashing at various places across the brain, it was eventually concluded that the clocks resided in the cerebral lobe area, and broadcast their timing information via pathways to the thoracic ganglia (Truman, 1974).

The clock for the fruit fly (*Drosophila*) activity rhythm is also in the brain. Two very bold experimentalists succeeded — to some extent — in transplanting the brain of this pest into its abdomen. Stop and reflect on just what that entailed: those scatologists among the readership will realize the brain is no larger than a fly frass, plus it is transparent; getting it into a tiny abdomnen is only slightly less difficult than getting it out of the even smaller head without decapitating the animal. But the neurosurgeons perservered and had success in four out of 55 tries. The design of the experiment was clever. As you should remember, the *per*^{S} mutant fly has a period of approximately 19 h in constant darkness and 24°C; the *per*^{O} mutant is arrhythmic. Brains from the former were transplanted into the latter, and when successful, the *per*^{O} mutants became rhythmic and their activity cycles described short periods. *Per*^{O} brain transplants into *per*^{O} flies were used as controls (Handler & Konopka, 1979). Cymborowski (1981) has demonstrated that the cricket (*Acheta domesticus*) brain functions similarly.

The first experiments attempting to locate the clock in the cockraod *Periplaneta americana*, indicated that its brain was also a hormonal pacemaker governing the animal's locomotory rhythm. That turned out to be incorrect, and to shorten a long story, it was eventually found that the clocks resided in the optic lobes and were not hormonal in action (Nishiitsutsuji-Uwo & Pittendrigh, 1968). This was confirmed in a most dramatic way by Terry Page (1982, 1983a,b) using the roach *Leucophaea maderae*. A knife cut between both of the optic lobes of the protocerebrum and the midbrain causes the persistent locomotor rhythm to cease (isolating jut one lobe does not). But if one waits 3–5 weeks, the rhythm returns, and its phase, when projected back to the day of surgery, was correlated with the phase of the preoperative rhythm, as if the optic oscillator had been running all the time. Histological examination of the tissue revealed that over the 3-5 week interval, nerve regeneration had rejoined the optic lobes to the midbrain (Page, 1983b). This finding set the stage for the next experiment.

If male cockroaches are raised from birth in 11 h of light alternating with 11 h of darkness, when their locomotor output is tested in constant darkness their average period is 22.7 h. For those raised under 13 h of light/13 h of darkness, their persistent rhythm has an average period of 24.2 h. The roughly 1-h difference between the two groups will endure for at least 5 months in constant conditions. So, to the experiment: The periods of individual animals from both groups were first verified and their optic lobes removed and replaced by lobes from the other group. After an interval of 26–58 days, regeneration of the neural pathways was completed and again the roaches became rhythmic. In each individual case the period shown was that of the donor optic lobe (Page, 1982; 1983b).

The two optic lobes are connected with each other via a neural pathway running across the midbrain. When this tract is cut it does not regenerate, eliminating all crosstalk such as mutual entrainment of the two ocular oscillators. In another clever experiment, Terry Page (1983a) first severed this tract in long-period animals, removed one optic lobe, and replaced it with a lobe from a short-period roach. Even when sufficient time had passed for nerve reattachment between the transplant and the midbrain, the activity rhythm, which had persisted the whole time after surgery (being governed by the intact clock), remained and expressed the same long period. One way of testing these animals to see if the transplants had connected, was to section the optic nerve of the previously untouched optic lobe. This done, the rhythmic expression switched to a short period: The new connection had been made, but the resident optic-lobe clock, even though the nerve pathway between the eyes had been cut, had somehow still suppressed the expression of the donor short-period clock. With time, however, the long-period lobe re-established its neural connection with the brain, and two things happened: (1) now it did not suppress the action of the short-day clock, and (2) it inserted a long-period into a roach's locomotor display. An animal now expressed *both* long and short periods! Figure 6-10, is a diagram I have made showing what an actogram containing such data looks like (to have represented the raw data here would have meant reducing them to an almost useless scrim of their former selves).

One last example will be given of bilateral, neurally based, optic-lobe associated, clocks — this one from a tenebroid beetle *Blaps gigas*. The changing sensitivity to light in this insect's eye was studied. A fine electrode inserted into the eye recorded its spike output (called the "on effect") when a flash of light was focused on the eye. When the measurement was repeated many times it was found, as seen in an electroretinogram (Fig. 6–11), that the magnitude of the response was 10–100 times greater at night, and that the difference persisted in constant conditions. The changing sensitivity of the eye — both eyes to be specific — was rhythmic.

There is also a distinct similarity with this rhythm and the tide-associated rhythms of many intertidal animals. When recordings are made simultaneously from both eyes, the light sensitivity rhythms of each eye of a few individuals spontaneously adopted different periods. Examples of this are seen in Figs 6-11 and 6-12. Tidal rhythms acting the same way are seen in Figs 3-8, 3-24, 3-25, 3-26, and 3-35.

In another experiment a 24-h light/dark cycle was presented to one eye through a light conducting fiber, while the other eye was kept in constant darkness. The rhythm of the former eye was entrained to the light/dark cycle, while the other assumed a long circadian period (Fig. 6-13) (Koehler & Fleissner, 1978). Note that the period of the latter was so long that the eyes crossed (so to speak). The two eye clocks controlling these rhythms are either not coupled to one another, or only weakly so.

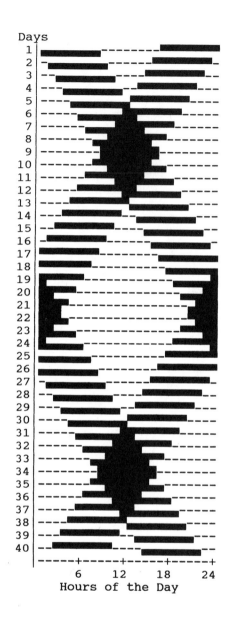

Figure 6-10 Diagrammatic representation of the activity pattern expressed by a cockroach that has a clock in one optic lobe that runs at a period of 25 h, while the one in the other lobe measures off 23-h intervals. The bars represent 8-h bouts of locomotor activity. In each hour during which the two rhythms cross, the amount of activity is doubled (based on data from Page, 1983b).

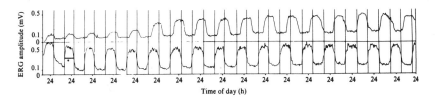

Figure 6-11 Electroretinograms of the right and left eyes of the beetle *Blaps gigas*. At intervals of 30 min the eyes were exposed to a 30 ms flash of light, and the changing amplitude of the "on-effect" recorded (it is greatest at night). In animals first tested in constant conditions the rhythms of the two eyes were in phase, but in some individuals, such as the one portrayed, over time the rhythms came out of synchrony (Koehler & Fleissner, 1978).

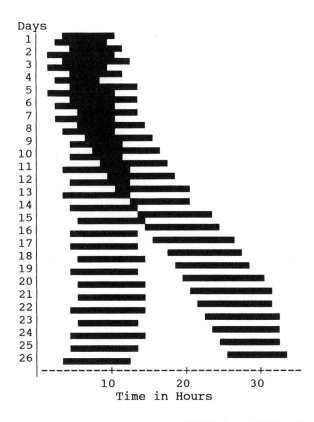

Figure 6-12 The spontaneous assumption of different period lengths, in constant conditions, of the circadian rhythms in light sensitivity of the left and right compound eyes of the beetle *Blaps gigas*. Each bar represents the maximum phase of each sensitivity rhythm (redrawn and modified from Koehler & Fleissner, 1978).

Time (in hours)

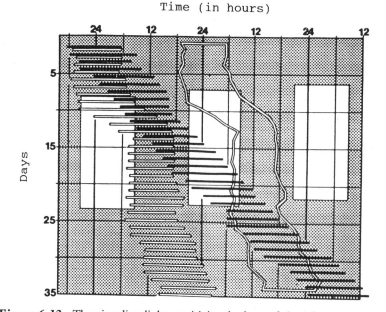

Figure 6-13 The circadian light-sensitivity rhythms of the left and right compound eyes of the beetle *Blaps gigas*. First notice that the two rhythms had assumed different periods by the third day in constant darkness (the dark time indicated by the stippling). On the ninth day the left eye (represented by the light bars) only was unilaterally offered a light/dark cycle (the light interval indicated by the unstippled box) and 6 days later had roughly entrained to the time of light off. The light/dark cycle was stopped on day 23, and the left eye's rhythm again assumed a circadian peiod. The right eye was kept in constant darkness the whole time. That the two ocular clocks are not coupled together is indicated by the unvarying slope of right eye's rhythmic response. The envelope of the night phases of the left eye was duplicated and reprinted to the right, showing that the two rhythms actually crossed one another between days 20 and 35 (Koehler & Fleissner, 1978).

Summary

Before we move on to the comparatively sparse knowledge of what is known about the clock driving tide-associated rhythms, I will rehash the above. Often there is more than one clock present, such as the paired SCNs, the paired ocular oscillators of *Aplysia* and *Bulla* and insects, the paired centers in the brain of silkmoths, and others not mentioned. These are usually pretty much redundant clocks, i.e., one can be removed without an individual's rhythm that is normally controlled by the two, changing much. Usually the two are coupled together, but in *Blaps* this certainly does not seem to be the case. That a clock — or clocks — is found to control one or more processes does not mean that it controls all of an individual's rhythms, as seen in silkmoths and fruit flies who use different clocks for eclosion and flight rhythms. And, returning to Chapter 1, some of these

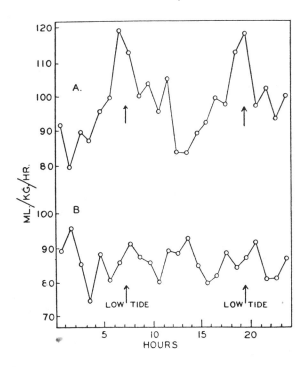

Figure 6-14 Oxygen consumption in the fiddler crab *Uca pugilator*. (A) form estimate of O_2 utilization of several crabs in constant dim illumination and 25°C. (B) The form estimate of several crabs in the same conditions after their eyestalks had been removed (Brown *et al.*, 1954).

clocks serve as master chronometers to co-ordinate lesser slave clocks in cells and organs. Thus, parts of a multicellular organism can be isolated in culture and continue to be rhythmic, where they are driven by freed slave timepieces. At a distant extreme, the unusually large single cell *Acetabularia*, that can live for months without its nucleus, can be cut into small fragments and each anucleate piece remains rhythmic (redundancy at its uttermost) (Mergenhagen & Schweiger, 1975). There is even evidence in this plant that there is a second clock that drives some of the plant's several rhythms (Schweiger *et al.*, 1986). Finally, some clocks function hormonally and others function neurally.

The Role of the Eyestalks

The Fiddler Crab (Uca)

As described in Chapter 3, changes in the rate of oxygen consumption are caused by the activity rhythm in the fiddler crab, *Uca pugilator*, so that respiration also takes on the form of a tide-associated rhythm (Fig. 6-14(A)). When the animal's eyestalks were removed, as seen in Fig.

6-14(B), the waveform in O_2 consumption was eliminated, meaning, presumably, that the locomotion rhythm was also destroyed (Edwards, 1950; Brown *et al.*, 1954). Paradoxically, although a pigment-dispersing hormone is known to be produced in the eyestalks, removal of the stalks of *U. pugilator* did not stop its color-change rhythm (Webb *et al.*, 1954; Fingerman & Yamamoto, 1964).

The Shore Crab (Carcinus)

Considerably more work has been done on the eyestalks of *C. maenas*. When the stalks are removed from this crab it becomes hyper-active and non-rhythmic (Naylor & Williams, 1968). You should remember from Chapter 3 that after a few days in the laboratory away from the tides this crab loses its rhythms, but exposure to one 15-h (or shorter) cold shock (1–4°C.) will restart the rhythm (Fig. 3-13) (Naylor, 1963). With this as background, arrhythmic intact crabs and stalkless crabs were subjected to a 4°C pulse: The cold shock caused only the former (the controls) to become rhythmic again. That the eyestalkless crabs remained arrhythmic suggested — because the only difference between the two groups were the missing eyestalks — that the clocks may have been removed.

The experiment was varied but still focused on the role of the eyestalks. Crabs were allowed to become arrhythmic — as a result of storage — and were then attached to the bottom of a vessel and submerged in 15°C water to a depth such that only the eyestalks rose, periscope-like, above the surface. Then 1°C seawater was slowly dripped — in ancient Chinese water-torture fashion — onto the eyestalks. This 10-h treatment re-initiated the activity rhythm (Williams & Naylor, 1967).

Crude eyestalk extracts were made and injected into crabs whose stalks had been removed, and who were thus undergoing heightened activity. This extract reduced the crabs' activity for 6–18 hours, after which it returned to its pre-injection level. The inhibition was found to be greatest in extracts prepared from eyestalks taken from crabs in their inactive phase (Naylor *et al.*, 1973). This result suggests, of course, a cyclic pattern of hormone release or synthesis, and thus the presence of an *endocrine* clock in the eyestalk — so the search began.

Within the eyestalk is located the X-organ/sinus gland complex (Smith & Naylor, 1972). A variety of hormones are known to be made in the X-organ (also known as Hanström's organ, or less colorfully as E_1) and stored and released by the sinus gland. One of these, called neurodepressing hormone — abbreviated to NDH — is the cause of the reduction in locomotor activity. Through dialysing, heating, and proteolytic digestion procedures it has been identified as a low molecular weight peptide (Aréchiga *et al.*, 1974). Although pure NDH is not yet available, it has been separated from many of the other components in a crude extract by column fractionation. The activity-reducing effect of partially-purified NDH from the Norway lobster *Nephrops norvegicus*, and the shore crab

Carcinus, have been compared on the latter. It produces approximately the same activity reduction in both species, the cross- sensitivity suggesting that they may be identical (Aréchiga *et al.*, 1979).

Next, the eyestalks were left attached, but holes were drilled through their covering exoskeleton and different parts of the peduncle destroyed by cautery. After recuperating from the selective searing, the crabs were shocked with a cold pulse, and examined over the next few days for rhythmicity. All except those whose X-organs had been ashed were rhythmic. The locomotor patterns of these exceptions were checked with periodogram analysis and arrhythmicity confirmed.

Using rhythmic crabs, eyestalks were removed during activity peaks or during troughs, and the optic peduncles fixed, stained with Aldehyde Fuchsin (it defines neurosecretory material), sectioned, and examined microscopically. One cell type in particular, during the inactive phase of the locomotor rhythm, was found to be loaded with neurosecretory material. This cell type is unique to the X-organ. Could this mean that the X-organ is a clock?

The penultimate approach was to remove the optic peduncles of rhythmic crabs, attempt to maintain them in culture, and, again using the Aldehyde Fuchsin technique, see if the NDH synthesis rhythm would persist. Although standard rough-culture procedures were used, most isolates survived only 2 or 3 days. Mainly because of culture problems, the results were not as clear-cut as one would desire, but there was a definite indication that most staining took place during the inactive stage (Williams *et al.*, 1979).

Lastly, rather than remove the eyestalks, holes were drilled through the exoskeleton and the optic tract cut between the cerebral ganglion and the base of the optic peduncle. It was reckoned that while this would *physically* isolate the peduncle from the rest of the animal, NDH might still be able to flow from the isolate into the animal via the hemolymph. First, as a precaution, just the exoskeleton was cut in 15 crabs; their rhythms persisted. Then, the optic tract of just one eye was severed. This caused heightened activity for 24–48 h followed by a return to the pre-sectioning level; the retention of a tidal rhythm could be seen in the records of some animals. A cold shock (4°C for 13 h) was then given and this caused a strong rhythm in all 22 subjects (as would be expected because one peduncle was left intact). That procedure was followed by the next logical variation, cutting the optic tracts of both eyestalks. This caused a much longer interval of hyperactivity — in a few cases up to the time of cold shock, 2 or more days later. After the low-temperature pulse, there was no indication of rhythmicity in the 36 crabs so mutilated (Williams, 1985).

The last finding in the above paragraph was not what was expected. Knowing that NDH plays an important role in producing a rhythmic display in locomotion, and that isolated stalks in culture suggested that the X-organ may be able to oscillate autonomously, why did the severed

Figure 6-15   A parabiotic pair of penultimate-hour crabs (*Sesarma reticulatum*) joined by openings through their carapaces. Before joining the two, the crab on top had a strong locomotor rhythm while the one on the bottom was arrhythmic. The upper crab's legs were caused to be cast off and the pair put in an actograph where locomotor activity of the lower crab was recorded. The experiment was designed to elucidate whether a blood-borne substance stimulating activity would be passed to the lower animal. It was not, the lower crab remained arrhythmic (Palmer, 1974).

optic peduncle, still within the eyestalk exoskeleton and bathed with hemolymph, not continue to function (even though the optic tract had been cut) and NDH flow to the body via the open circulatory system? An obvious answer might be that the severed optic peduncles survived no better *in situ* than they did in culture — it was at least 2 days after the operation that the cold shock was applied. Another interpretation could be that the release of NDH from the X-organ/sinus gland complex is brought about by a clock located elsewhere in the body, say in the central nervous system. That clock would send its commands to the eyestalk to release NDH via either a neurohumor transferred axonally, or by action potentials, along the optic track. It may be the latter: Electrical stimulation is known to cause sinus-gland hormone release in the crayfish (Aréchiga *et al.*, 1977). Cutting the optic tract in that animal precludes the messages being delivered. But how does one explain rhythmicity being re-started by chilling just the eyestalks?

That the basic clock(s) may exert its influence electrically, rather than hormonally, is supported by experiments run on the penultimate-hour crab, *Sesarma*. Crude stalk extracts were made from rhythmic crabs at peak and at minimal activity times, and injected into crabs made arrhythmic by storage. No consistently reproducible results were obtained. To test for blood-borne hormones made *anywhere* in a crab's body was accomplished by parabiosing a rhythmic crab to an arrhythmic one. Complementary holes were made in the dorsal carapace exoskeleton to one side of the midline and the crabs conjoined with sealing wax, hole to hole, so their hemolymphs mixed easily. The rhythmic crab was caused to cast off all 10 of its legs, and each funny-looking pair (Fig. 6–15), bonded for life, were placed in their own actograph. Without legs on the piggy-backed crab, the only activity that could be recorded was that of the supporting one below. That a sans-legs crab continued to remain alive could be easily checked any time by watching for eyestalk movements. Not one case of 47 combinants ever showed a sign of even a wonky rhythm. Control pairs, consisting of parabiosed rhythmic animals, described fine rhythms, indicating that the procedure was not as drastic as one might think from just looking at the laboratory made chimeras (Palmer, 1974).

Other Crustacean Decapods

I will digress for a few paragraphs from the tidal rhythms of marine crabs, to the control of the circadian rhythms of freshwater crayfish. The logic behind this "wrong turn" is simple: the greatest amount of work on the clocks of crustaceans has come from studies on the daily rhythms of crayfish. What has been learned from these studies should be useful as a beacon to illuminate what may be going on in marine crabs.

Two main organs in crayfish have been found to contain clocks: the optic peduncle of eyestalks and the supraesophageal ganglion (the "brain").

A variety of circadian rhythms has been found in several species of crayfish (for a list, see the fine review by Aréchiga *et al.*, 1993; but here, as the topic is not part of the main thrust of this monograph, I will attribute all the studies to just the generic "crayfish"). Let us begin with the optic peduncle. Hans Kalmus (1938) was probably the first investigator to demonstrate that removal of the eyestalks destroyed a crayfish's locomotor rhythm. This finding was confirmed by several others and became part of chronobiological dogma until Page and Larimer (1972) found that the rhythm in some crayfish would persist after eyestalk ablation. The photoreceptors for the entrainment of the locomotory rhythm are not in the retina, but rather are found in the supraesophageal ganglion (Page & Larimer, 1972). Studies then drifted toward another crayfish rhythm — the changing visual sensitivity of the eye (a rhythm that is also found in *Carcinus* (Aréchiga *et al.*, 1974)).

The compound eyes of crustaceans are made up of many individual photoreceptors called ommatadia. In the bases of the ommitidia are proximal-pigment granules that migrate between two positions: they are dispersed within the retinula during the daytime, and are concentrated at the axonal ends of the ommatidia at night (Fig. 6-16). This migration rhythm was first shown to persist in constant darkness (only) by Bennitt (1932). Around each ommatidium is a set of slender cells that contain the so-called distal pigment granules. That shielding pigment is dispersed

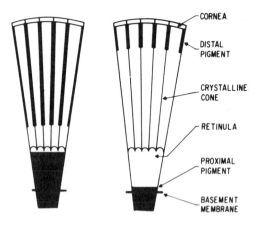

Figure 6-16 A diagrammatic representation of the two extremes of pigment migration in the compound eyes of a crustacean. Each ommatidium consists of a cornea, crystalline cone and retinula. On the left is seen the distribution of the distal and proximal pigments during daylight; on the right the arrangement at night (modified from Aréchiga *et al.*, 1993).

throughout the length of these cells during the daylight, preventing stray light from being scattered between the ommatidia. At night this pigment is concentrated at the distal end of each ommatidium. The daily migrations of this pigment also persist as a circadian rhythm in constant light (only) (Welsh, 1930). The migrations of these two pigments produce a light-responsiveness rhythm that can be easily recorded in an electroretinogram.

The movements of the distal pigment does not respond directly to light; instead, it is controlled hormonally: an 18-residue peptide called pigment-dispersing hormone (PDH) is released under illumination; it promotes the dispersion of the distal pigments. A red pigment-concentrating hormone (RPCH), an eight-residue peptide, causes retraction of the pigment. Antibodies made against these peptides have been produced and then used to locate the cell bodies that synthesize them. RPCH, for instance, is made in about 20 neurosecretory cells of the X-organ (Mangerich *et al.*, 1986). It, like PDH, is stored in the sinus gland.

This light-responsiveness rhythm will persist in cultured eyestalks (Sánchez & Fuentes-Pardo, 1977) and even in isolated retinas (Noguerón & Aréchiga, 1987) (this rhythm, and some accompanying ultradian rhythms, are all lengthened by deuterium oxide administration (Pardo & Sáenz, 1988)). The proximal-pigment position rhythm persists with a reduced amplitude, but the distal-pigment migration cycle does not, which is curious because at least one of the driving hormones, RPCH, continues to be secreted rhythmically (Rodríguez-Sosa *et al.*, 1990). Oddly, the light receptor for the entrainment of the electroretinogram-amplitude rhythm is in the sixth abdominal ganglion (Fuentes-Pardo & Inclán-Rubio, 1987).

The optic peduncles are connected to the supraesophageal ganglion (SOG) by the optic nerves. The synchronization of the independent clocks in the eyestalks could be controlled by neurosecretions passed from one to the other via the hemolymph, or neuronally via the optic nerves and the SOG (an abbreviationist's nightmare: it makes one think of a donut dunked too long in breakfast coffee). Nerve pathways for this route have been found, and a longitudinal bisection of the supraesophageal ganglion causes a loss in phase-coupling of the two retina (Barrera-Mera, 1976). The SOG also may have clock properties of its own: Page and Larimer (1975) severed the descending pathways from the SOG to the thoracic ganglia (where the motor neurons to the legs reside) and destroyed the locomotor rhythm. Cutting the optic nerves, or removing the SOG, annihilated the distal-pigment rhythm of the retina.

Finally, there appears to be a clock in the abdominal ganglia also. Abdominal ganglia 2–6 in culture under constant darkness and temperature display a circadian spike output rhythm (Block, 1976). Motor neurons contained there are known to respond directly to illumination (Edwards, 1984), but it is not known if they entrain the abdominal-ganglia spike rhythm.

The Amphipod *(Corophium)*

An ingenious method was devised to locate just where the clock(s) governing the swimming rhythm in this small crustacean is located. Remember, as described in Chapter 4, that as strange as it might seem, short-duration sub-zero temperature pulses do not freeze this tiny critter to death, they rephase its rhythm. Remember also that what may be a clock in *Carcinus* was first located by dripping ice water on its eyestalks (Williams & Naylor, 1967). Doing something as easy as that is impossible in *Corophium* for two reasons: the animal is so small that a drop to it must be like a tsunami is to Japan, and it does not have stalked eyes. Thus, the following technique was devised. A small raft was made of thin Styrofoam and several animals lashed to it with thread (Fig. 6-17). The raft was floated on ethylene glycol cooled to below freezing. A thin copper wire dangling in the cold bath and pushed up through the Styrofoam was chilled by the ethylene glycol. The wire could be wrapped about specific parts of an amphipod body chilling touched points to below freezing. This was either the supra- or sub-esophageal ganglion, the middle of an animal's body, or the posterior end of each subject could be chilled.

A −3°C pulse, lasting 3 h and given at the expected time of high tide, was followed by the release of the animals into 12.5°C seawater where the timing of their swimming behavior could be recorded. The results of one such experiment, this one focusing on the esophageal ganglia, is seen in Fig. 6-18. The group at the top of the figure (A) are simply freshly collected

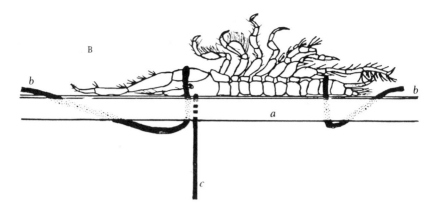

Figure 6-17 The means of chilling small areas of the tiny amphipod, *Corophium*. Captive animals were lashed with cotton thread (b), to a Styrofoam raft (a), through which a copper wire (c) was threaded. The wire touched a specific part of a subject's body and drained the heat from it down into the sub-zero ethylene glycol on which the raft floated (Harris & Morgan, 1984b).

animals. Those in (B) were bound to a raft but were not chilled. In the last two cases the animals were subjected to 3 h at −3°C and then moved to constant conditions. Those in (C) had their entire bodies chilled, while those in (D) had just their esophageal ganglia chilled. After all combinations of body-part chilling were performed, it was found that chilling the supra- and sub-esophageal ganglia (or whole-body chilling because, of course, it included the brain) caused phase delays, while only midbody or posterior end chilling did not.

Cognizant of the pacemaking role of the eyestalks in *Carcinus*, just for good measure the eyes of *Corophium* were removed. A transfer needle was heated to the temperature of a red-hot poker and the eyes burned out of 10 animals. In other animals the eyes were untouched, but the area directly behind them was cauterized. This treatment had no effect on their rhythm.

Thus, a clock, or clocks if the supra- and sub-esophageal ganglia can function separately — or maybe just the phase-setting machinery if it is separate from the clock — certainly appears to reside in the head of *Corophium* (Harris & Morgan, 1984b).

Endocrine Control in a Polychaete

The persistent monthly rhythm in the syllid *Typosyllis prolifera* was discussed in Chapter 5. During the last half of each month in the summer, the posterior end of these worms metamorphoses into a "stolon," the reproductive rendition of the animal. This transformation is under the control of two endocrine-producing regions in the body: the prostomium

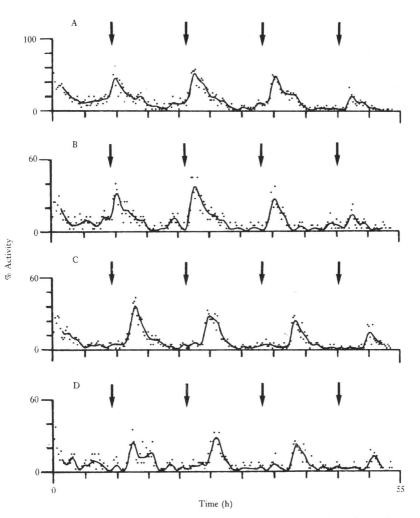

Figure 6-18 Delay of phase caused by chilling the supra-esophageal ganglion of *Corophium*. (A) the swimming of untreated freshly collected animals. (B) the rhythm of amphipods bound to a floating raft, but not subjected to sub-zero temperatures. (C) as in (B) but totally immersed in −3°C for 3 h before the swimming rhythm was tested. (D) only the supra-esophageal ganglia of these animals were chilled to −3°C for 3 h before being tested. Individual points represent the mean percentage swimming every 12 min. The data were tarted up using a 15-point moving average. The rain of arrows signifies the times of expected high tides (Harris & Morgan, 1984b).

(that looks like the first anterior segment, but is technically not a segment), and the proventriculus (see Fig. 5-3 for location). Secretions from the former stimulate stolonization, and the latter inhibit it. Remember that reproduction in this species is under photoperiodic control: they are long-day animals. Cultures kept in the laboratory under short days do not form stolons. But, if prostomia from animals maintained in long days are implanted into short-day animals, the recipients are induced to transform their posterior ends into functional stolons. Making such a transplant just causes one episode of stolon formation and release, then the animal returns to its customary non-reproductive winter life.

Or, if the proventriculi are excised from animals maintained in short days, stolons will form because the proventriculus plays an inhibitory role. And, in its absence, one stolon after another is produced without an interval of re-growth in between (Fig. 5-3). Soon almost the whole worm has been transformed into stolons!

Whether the clock that controls this animal's monthly rhythm is part of that endocrine system, or is separate and just controls the system's secretions is not known (Franke, 1986).

The Clock Hypotheses

The Circatidal Clock Hypothesis

This idea is the older, traditional hypothesis: The clock controlling tidal rhythms is envisioned as running at a rate approximating 12.4 h, thus matching the average period of most tides. The hypothesis has been re-championed most recently by Reid and Naylor (1989; 1993) who used it to explain the results of their "hypo-osmotic shock" experiments on *Carcinus* (Chapter 4).

When Williams (1991) chose to study the rhythms of *Carcinus*, however, she got results that do not fit with the above idea. In constant conditions only one third of her sample (18 of 54 crabs) produced traditional tidal rhythms, i.e., two peaks per lunar day, and in most cases of this small fraction, one of the peaks was only a diminutive sibling of the other and thus barely discernible and short-lived. In the other animals only the expected night-time peak, or the expected daytime peak, was displayed (compare Fig. 3-15 with Fig. 3-17).

The Circalunidian Clock Hypothesis

As introduced in Chapter 3, this hypothesis suggests that what seems axiomatic is not, i.e., the basic period of an organismic tidal rhythm does not match the 12.4-h period of the tides. This non-intuitive, heretical conclusion first arose from finding totally unexpected results, gained by studying individual rather than group responses. The most compelling find was that the twice-daily peaks of a tidal rhythm could scan the solar day at different rates (Figs 3-8, 3-24, 3-25, 3-26, 3-35), i.e., they could each independently adopt their own period. Additionally, one peak might disappear, either temporarily or permanently, while the other remained intact (Figs 3-2, 3-5, 3-6, 4-17, etc.); one peak might split (Figs 3-4, 3-5, etc.) while the other remained unchanged; the forms and peak heights of the two sometimes alternated aperiodically (Fig. 3-7); and certain chemicals will sometimes alter the period, or annihilate one of the peaks, but not the other (Fig. 6-5). All, most, or some of these same properties have now been seen in seven crab species belonging to five genera (Palmer & Williams, 1986, 1987, 1993; Palmer, 1988, 1989), in an isopod (Klapow, 1972; Enright, 1976), in littoral snails (Petpiroon & Morgan, 1983), in a clam (Williams *et al.*, 1993), and in fishes (Green, 1971; Northcott *et al.*, 1990). In short, the two daily peaks can act quite independently of one another. To explain this curiosity we (Palmer & Williams, 1986; and elaborated by Palmer, 1990b) postulated that each peak must be controlled by its own clock, and that the basic period of each of the timepieces was 24.8 h. We further postulated that the two independent clocks' outputs were strongly coupled (as indicated by, for instance, the tenacity displayed during the zigzag seen in Fig. 3-3–together 180° out of phase (Fig. 3-9); thus while phase locked the overt display was indistinguishable from a tidal or circatidal rhythm. But when the coupling breaks, and/or the mutual enhancement between clocks is lost, as sometimes happens in the laboratory, then the dual nature of the clock control is made manifest.

This twin-clock epiphany adequately explains results obtained by Barnwell (1968) in a study of the translocation of two species of *Uca* (*mordax* and *minax*) from the Caribbean coast to the Pacific coast of Costa Rica. On the former shoreline there is only one tide/lunar day, and that is the pattern shown in crabs when tested in the constancy of the laboratory. But after transporting them across that beautiful country and exposing them in cages for a week to the twice/day Pacific tides, that frequency became the waveform that the animals described in the laboratory. Exposure to the semidiurnal tides thus called the second clock into play. The same conclusion can be drawn from Green's work (1971). He started with fish (a cottid, *Oligocottus maculosus*) made arrhythmic by storage in the laboratory. Then he placed them for a few days in cages in two tidal pools, one low on the shoreline and thus exposed to two tides per lunar day, and the other higher up on the shore and thus reached only once a day by the semidiurnal, unequal tides. The timing of the different exposure patterns was adopted by the fish, starting one or both of the lunidian clocks. Northcott *et al.* (1991) found a suggestion of this with another fish, and Williams (1995) has, under some experimental conditions, repeated the finding in clams (Fig. 4-26, p.122). The hypothesis provides a fine explanation for the periodic disappearance of the night-time peak in the *Hantzschia* vertical migration rhythm.

A different kind of logic can be applied to the feasibility of a circalunidian clock hypothesis. Remember that the main function of a clock is to alert an organism *in advance* of the coming of some periodic environmental event, such as the time of the next low tide. How useful a horologue is to its owner depends on two factors: the accuracy of the clock, and the precision of the periodic event that the clock is trying to predict. For instance, depending on the airline that one flies with, using a slow-running wrist watch can result in missed flights. As pointed out in the first Chapter and Fig. 1-6 (p.11), the peak-to-peak punctuality of the tides (especially on the east coast of the United States) is even worse than the airlines I normally use. In fact, the tides are so erratic that it is difficult to believe that natural selection would have played a role in selecting a clock with a basic 12.4-h period: Examining the upper entry on Fig. 1-6, you can see that at some times of a month successive tidal periods can vary by almost 3 h, making them very elusive temporal targets for a clock to hit! However, look at the bottom exhibition in Fig. 1-6, which describes the variation around a basic period of 24.8 h, the interval of one low tide and its repeat the next day. There are still variations associated with the phases of the moon, but the mean deviation from 24.8 h for the month is only 5.7 min — a scant 10% of the average scatter seen in (A) of Fig. 1-6. Clearly, for this reason, a basic lunidian clock would be much more useful to an organism, and thus a likely choice for natural selection. Still, however, the magnitude of even this nonuniformity precludes evolutionary pressure for a highly accurate clock. Thus, tide-associated rhythms are notoriously imprecise.

If the clock that controls tidal rhythms really has a basic period approximating 24.8 h, would natural selection have created a second clock that only runs approximately 3% faster to control solar-day rhythms? Begging the question further, the circa periods of both solar- and lunar-day rhythm are known to overlap widely. In attempting to get at the answer to this challenge, comparing the properties of solar-day and lunar-day rhythms is of limited help: It makes little difference that both clockworks are only minutely influenced by different constant temperatures or by a whole variety of chemicals in the environment; every clock, even the one on your wrist, must have the above properties to be of any use as a timepiece. Evolution would be expected to have produced clocks with these characteristics even if each phylum arose independently as a result of spontaneous generation out of the muck. It is thus more useful to compare *non* adaptive aspects of the two rhythms since Mother Nature would not have selected for them. We could use the foibles expressed in the laboratory where the unnaturalness of constant conditions puts great stress on the timepieces and on occasion brings out the worst in them.

The failings of the clock, or clocks, that controls lunar- and solar-day rhythms are identical: both change period spontaneously in the laboratory; both are known to be influenced by the administration of deuterium oxide, alcohol, and azadirachtin; both allow splitting of peaks (Wiedenmann, 1983); and both either stop temporarily, or become uncoupled from the processes that they are causing to be rhythmic (Chapter 3). This concatenation of shared flaws (all undesirable, and thus, as dictated by dogma, traits that would not have been evolutionarily instilled) support a hypothesis that the same clock drives both solar-day and lunar-day rhythms.

Both solar- and lunar-day clocks are associated with ultradian sub-cycles. As pointed out earlier, a popular explanation promulgated to explain the timing of *circadian* rhythms invokes the coupling of many short-period oscillators, whose existence is hypothetically demonstrated by ultradian rhythms expressed by unruly individual oscillators that have broken free of the conglomerate. Now ultradian rhythms have been found accompanying tide-associated rhythms also. According to the meta-oscillator hypothesis, U-oscillators would only have to be coupled in a slightly different combination and fundamental circalunidian rhythms would result. Easy to do on paper, is it not? The validity of that speculation awaits the further elucidation of cellular mechanisms.

Anatomical evidence, coming from invertebrates, supports a multiple clock hypothesis. Many animals have been demonstrated to have at least two master clocks, e.g., the green crab may have one in each eyestalk, as do the eyes of mollusks such as *Aplysia* and *Bulla*, and the eyes of many insect species. And while the clocks in these invertebrates are known to be physically or hormonally coupled together; it is also known that the coupling can be broken experimentally or spontaneously (Figs 6-10, 6-13). Furthermore, many organisms that express tide-associated rhythms also display solar-day rhythms. Intertidal crabs, such as *Uca*, *Sesarma*, and *Carcinus*, have both tidal activity rhythms and solar-daily rhythms in color change and/or phototaxis. The locomotion rhythms of *Sesarma*, *Carcinus*, *Hantzschia* and *Cyclograpsus* also contain a daily component (Chapter 3). Again, economy of hypothesis would suggest multiple functions of a single clock type.

There is more evidence for multiple clocks. Individual organisms contain many copies of the same clock within their bodies as is easily demonstrated by extirpating parts of whole organisms, maintaining the isolates in culture, and finding that they continue to display the same rhythms as before their excision from an unwilling donor. Even *uni*cellular organisms such as *Acetabularia* possess clock replicates within their cytoplasm (Mergenhagen & Schweiger, 1975). It is also known that the solar-day clocks within a single organism can run at different rates (e.g., the sleep/wakefulness and body temperature rhythms of humans often adopt different circa periods in constant conditions (Wever, 1979)), as do the ERG rhythms in *Blaps* (Figs 6-11, 6-12 & 6-13), activity in the cockroach (Fig. 6-10), photosynthesis and stomate opening in the bean (Hennessey & Field, 1992), the eyes of *Bulla* (Page & Nalovic, 1992), and even the bioluminescent rhythms in the unicell *Gonyaulax* (von der Heyde *et al.*, 1992; Roenneberg and Morse, 1993) — to mention just a few.

There is one piece of the single-clock speculation type that does not easily fit, and that is the fact that light/dark cycles, the strongest entraining agent known for solar-day rhythms, have no synchronizing clout with tidal rhythms. That lack is, of course, an absolute necessity, for if day/night alterations did entrain tidal rhythms they would no longer be tidal rhythms. The answer to this enigma may be that the mechanism that adjusts the clock to environmental cycles is separate from the timekeeping mechanism.

Maybe it is a function of the coupling mechanism; perhaps when natural selection chose a transmuting coupler for lunar-day rhythms, it also chose one programed to ignore day/night cycles. If the coupler did not have this immunity it would be of no use to its proprietor.

Literature Cited

Aréchiga, H., Huberman, A. and Naylor, E. 1974. Hormonal modulation of circadian neural activity in *Carcinus maenas*. *Proc. R. Soc., Lond.*, 187B: 299–313.

Aréchiga, H., Huberman, A. and Martinez-Palomo, A. 1977. Release of neurodepressing hormone from the crustacean sinus gland. *Brain Res.*, 128: 93–108.

Aréchiga, H., Williams, J.A., Pullin, R.S. and Naylor, E. 1979. Cross-sensitivity to neuro-depressing hormone and its effect on locomotor rhythmicity in two different groups of crustaceans. *Gen. Comp. Endochr.*, 37: 350–357.

Aréchiga, H., Fernández-Quiróz, A., Fernández de Miguel, F. and Rodríguez-Sosa, L. 1993. The circadian system of crustaceans. *Chronobiol. Int.*, 10: 1–19.

Bargiello, T.A., Jackson, F.R. and Young, M.W. 1984. Restoration of circadian behavioural rhythms by gene transfer in *Drosophila*. *Nature*, 312: 752–754.

Barnwell, F.H. 1968. The role of rhythmic systems in the adaptation of fiddler crabs to the intertidal zone. *Am. Zool.*, 8: 569–583.

Barrera-Mera, B. 1976. Effect of cerebroid ganglion lesions on ERG circadian rhythm in the crayfish. *Physiol. Behav.*, 17: 59–64.

Bennitt, R. 1932. Diurnal rhythm of the distal pigment cells in the eyes of the crayfish retina. *Physiol. Zool.*, 5: 65–69.

Block, G.D. 1976. Evidence for an entrainable circadian oscillator in the abdominal ganglia of crayfish. *Neurosci. Absts.* 2: 315.

Block, G.D. and McMahon, D.G. 1984. Cellular analysis of the *Bulla* ocular pacemaker system: localization of the circadian pacemaker. *J. Comp. Physiol.*, 155: 387–395.

Block, G.D., Khalsa, S.B., McMahon, D.G., Michel, S. and Guesz, M. 1993. Biological clocks in the retina: cellular mechanisms of biological timekeeping. *Int. Rev. Cytol.*, 146: 83–143.

Brown, F.A., Bennett, M.F. and Webb, H.M. 1954. Daily and tidal rhythm of O_2-consumption in fiddler crabs. *J. Cell. Comp. Physiol.*, 44: 477–506.

Bruce, V.G. and Pittendrigh, C.S. 1960. An effect of heavy water on the phase and period of the circadian rhythm in *Euglena*. *J. Cell. Comp. Physiol.*, 56: 25–31.

Citri, Y., Colot, H.V., Jacquier, A.C., Yu, Q., Hall, J.C., Baltomore, D. and Rosbash, M. 1987. A family of unusually spliced biologically active transcripts encoded by a *Drosophila* clock gene. *Nature*, 326: 42–47.

Cymborowski, B. 1981. Transplantation of circadian pacemaker in the house cricket, *Acheta domesticus*. *J. Interdiscip. Cycle Res.*, 12: 133–140.

DeCoursey, P.J. and Buggy, J. 1989. Circadian rhythmicity after neural transplant to hamster third ventricle: Specificity of suprachiasmatic nuclei. *Brain Res.*, 500: 263–275.

Dowse, H.B. and Palmer, J.D. 1972. The chronomutagenic effect of deuterium oxide on the period and entrainment of a biological rhythm. *Biol. Bull.*, 143: 513–524.

Dowse, H.B. and Ringo, J. 1987. Further evidence that the circadian clock in *Drosophila* is a population of coupled ultradian oscillators. *J. Biol. Rhythms*, 2: 65–76.

Dowse, H.B. and Ringo, J. 1992a. Do ultradian oscillators underlie the circadian clock in *Drosophila*? In: Lloyd, D. and Rossi, E.L. (Eds), *Ultradian Rhythms in Life Processes: An Inquiry into Fundamental Principles of Chronobiology and Psychobiology*, pp. 105–115. Springer-Verlag, London.

Dowse, H.B. and Ringo, J. 1992b. Is the circadian clock a "meta-oscillator"? Evidence from studies of ultradian rhythms in *Drosophila*. In: Young, M. (Ed.), *Molecular Genetics of Biological Rhythms*, pp. 197–220. Marcel Dekker, New York.

Dunlap, J.C. 1993. Genetic analysis of circadian clocks. *Ann. Rev. Physiol.*, 55: 683–728.

Edmunds, L.N. 1988. *Cellular and Molecular Bases of Biological Clocks*. Springer-Verlag, New York.

Edwards, D.H. 1984. Crayfish extraretinal photoreception. I. Behavioural and motor neuron responses to abdominal illumination. *J. Exp. Biol.*, 109: 291–306.

Edwards, G.A. 1950. The influence of eyestalk removal on the metabolism of the fiddler crab. *Physiologia Comp. Oecol.*, 2: 34–50.

Enright, J.T. 1971a. Heavy water slows biological timing processes. *Z. vergl. Physiol.*, 75: 1–16.

Enright, J.T. 1971b. The internal clock of drunken isopods. *Z. vergl. Physiol.*, 75: 332–346.

Enright, J.T. 1976. Plasticity in an isopod's clockworks: shaking shapes form and affects phase and frequency. *J. Comp. Physiol.*, 107: 13–37.

Eskin, A. 1971. Properties of the *Aplysia* visual system. *In vitro* entrainment of the circadian rhythm and centrifugal regulation of the eye. *Z. vergl. Physiol.*, 74: 353–371.

Eskin, A., Takahashi, J.S., Zatz, M. and Block, G.D. 1984. Cyclic guanosine 3′,5′-monophosphate mimics the effects of light on a circadian pacemaker in the eye of *Aplysia*. *J. Neurosci.*, 4: 2466–2471.

Ewer, J., Frisch, F., Hamblen-Coyle, M.J., Rosbash, M. and Hall, J. C. 1992. Expression of the *period* clock gene within different cell types in the brain of *Drosophila* adults and mosaic analysis of these cells' influence on circadian behavioral rhythms. *J. Neurosci.*, 12: 3321–3349.

Feldman, J.F. 1967. Lengthening the period of a biological clock in *Euglena* by cycloheximide, an inhibitor of protein synthesis. *Proc. Natl. Acad. Sci.*, 57: 1080–1087.

Feldman, J.F. and Hoyle, M. 1973. Isolation of circadian clock mutants of *Neurospora crassa*. *Genetics*, 75: 605–613.

Fingerman, M. and Yamamoto, Y. 1964. Daily rhythm of color change in eyestalkless fiddler crabs, *Uca pugilator*. *Amer. Zool.*, 4: 334.

Franke, H.-D. 1986. The role of light and endogenous factors in the timing of the reproductive cycle of *Typosyllis prolifera* and some other polychaetes. *Am. Zool.*, 26: 433–445.

Fuentes-Pardo, B. and Inclán-Rubio, V. 1987. Caudal photoreceptors synchronize circadian rhythms in crayfish. I. Synchronization of ERG and locomotor circadian rhythm. *Comp. Biochem. Physiol.*, 86A: 523–527.

Garcia, E.S. and Rembold, H. 1984. Effects of azadirachtin on ecdysis of *Rhodnius prolixus*. *J. Insect Physiol.* 30: 939–941.

Gaston, S. and Menaker, M. 1968. Pineal function: the biological clock in the sparrow? *Science*, 160: 1125–1127.

Goodenough, J.E. 1978. The lack of effect of deuterium oxide on the period and phase of the monthly orientation rhythm in planarians. *Int. J. Chronobiol.*, 5: 465–476.

Green, J.M. 1971. Field and laboratory activity patterns of the tidepool cottid, *Oligocottus maculosus*. *Can. J. Zool.*, 49: 255–265.

Han, S.-Z. and Englemann, W. 1987. Azadirachtin affects the circadian rhythm of locomotion in *Leucophaea maderae*. *J. Interdiscipl. Cycle Res.*, 20: 71–79.

Handler, A.M. and Konopka, R.J. 1979. Transplantation of a circadian pacemaker in *Drosophila*. *Nature*, 279: 236–238.

Hardin, P.E., Hall, J.C. and Rosbash, M. 1990. Feedback of the *Drosophila* period gene product on circadian cycling of its messenger RNA levels. *Nature*, 343: 536–540.

Harris, G.J. and Morgan, E. 1984a. The effects of ethanol, valinomycin and cycloheximide on the endogenous circa-tidal rhythm of the estuarine amphipod *Corophium volutator*. *Mar. Behav. Physiol.*, 10: 219–233.

Harris, G.J. and Morgan, E. 1984b. The location of circa-tidal pacemakers in the estuarine amphipod *Corophium volutator* using a selective chilling technique. *J. Exp Biol.*, 110: 125–142.

Hayes, C.J. and Palmer, J.D., 1976. The chronomutagenic effect of deuterium oxide on the period and entrainment of a biological rhythm: the reestablishment of lost entrainment by artificial LD cycles. *Inter. J. Chron.*, 4: 63–69.

Hennessey, T.L. and Field, C.B. 1992. Evidence of multiple circadian oscillators in bean plants. *J. Biol. Rhythms*, 7: 105–113.

Jacklet, J.W. 1969. Circadian rhythm of optic nerve impulses recorded in darkness from the isolated eye of *Aplysia*. *Science* 164: 562–563.

Jacklet, J.W. 1985. Neurobiology of circadian rhythm generators. *Trends Neurosci.*, 8: 69–73.

Jacklet, J.W. and Geronimo, J. 1971. Circadian rhythm: population of interacting neurons. *Science*, 174: 299–302.

Kalmus, H. 1938. Das Aktogram des Flusskrebs und seine Beeinflusung durch Organextrakte. *Z. vergl. Physiol.*, 25: 798–802.

Keller, S. 1960. Über die Wirkung chemischer Faktoren auf die tagesperiodischen Blattbewegungen von *Phaseolus multiflorus*. *Z. Bot.*, 48: 32–57.

Klapow, L.A. 1972. Natural and artificial rephasing of a tidal rhythm. *J. Comp. Physiol.*, 79: 233–258.

Koehler, W.K. and Fleissner, G. 1978. Internal desynchronisation of bilateral organised circadian oscillators in the visual system of insects. *Nature*, 274: 708–710.

Konopka, F.J. and Benzer, S. 1971. Clock mutants of *Drosophila melanogaster*. *Proc. Natl Acad. Sci.*, 68: 2112–2116.

Kupfermann, I. 1967. A circadian locomotion rhythm in *Aplysia california*. *Physiol. Behavior*, 3: 179–182.

Lehman, M.N., Silver, R., Gladstone, W.R., Kahn, R.M., Gibson, M. and Bittman, E.L. 1987. Circadian rhythm restored by neural transplant: immunochemical characterization of graft and its integration with the host brain. *J. Neurosci.*, 7: 1626–1638.

Lickey, M., Wozniak, J., Bock, G., Hudson, D. and Augter, G. 1977. The consequences of eye removal for the circadian rhythm of behavioral activity in *Aplysia*. *J. Comp. Physiol.*, 118: 121–143.

Liu, W., Zwiebel, L.J., Hinton, D., Benzer, S., Hall, J.C. and Rosbash, M. 1992. The *period* gene encodes a predominantly nuclear protein in adult *Drosophila*. *J. Neurosci.*, 12: 2735–2744.

Li-Weber, M., de Groot, E.J. and Schweiger, H.-G. 1987. Sequence homology to the *Drosophila per* locus in higher plant nuclei and in *Acetabularia* chloroplasts. *Mol. Gen. Genet.*, 209: 1–7.

Mangerich, S., Rainer, K. and Dircksen, H. 1986. Immunocytochemical identification of structures containing putative red-pigment-concentrating hormone in two species of decapod crustaceans. *Cell. Tissue Res.*, 245: 377–386.

McDaniel, M., Sulzman, F.M. and Hastings, J.W. 1974. Heavy water slows the *Gonyaulax* clock: A test of the hypothesis that D_2O affects circadian oscillations by diminishing the apparent temperature. *Proc. Nat. Acad. Sci.*, 71: 4389–4391.

Mergenhagen, D. and Schweiger, H.-G. 1975. Circadian rhythm in oxygen evolution in cell fragments of *Acetabularia mediterranea*. *Exp. Cell Res.*, 92: 127–130.

Michel, S., Geusz, M.E., Zaritsky, J.J. and Block, G.D. 1993. Circadian rhythm in membrane conductance expressed in isolated neurons. *Science*, 259: 239–241.

Moore, R.Y. and Eichler, V.B. 1972. Loss of circadian adrenal corticosterone rhythm following suprachiasmatic lesion in rat. *Brain Res.*, 42: 201–206.

Naylor, E. 1963. Temperature relationships of the locomotor rhythm of *Carcinus*. *J. Exp. Biol.*, 49: 669–679.

Naylor, E. and Williams, B.G. 1968. Effects of eyestalk removal on rhythmic locomotor activity in *Carcinus*. *J. Exp. Biol.*, 49: 107–116.

Naylor, E., Smith, G. and Williams, B.G. 1973. Role of the eyestalk in the tidal activity rhythm of the shore crab *Carcinus maenas*. In: J. Salanki (Ed.), *Neurobiology of Invertebrates*, pp. 423–429. Hungarian Academy of Sciences, Budapest.

Nishiitsutsuji-Uwo, J. and Pittendrigh, C.S. 1968. Central nervous system control of circadian rhythmicity in the cockroach. III. The optic lobes, locus of the driving oscillation? *Z. vergl. Physiol.*, 58: 14–46.

Noguerón, I. and Aréchiga, H. 1987. Ritmo circádico de sensibilidad a la luz en la retina aislada del acocil. *Bol. Estud. Med. Biol. Mex.*, 35: 165.

Northcott, S.J., Gibson, R.N. and Morgan, E. 1990. The persistence and modulation of endogenous circatidal rhythmicity in *Lipophrys pholis*. *J. Mar. Biol. Ass. UK*, 7: 815–827.

Northcott, S.G., Gibson, R.N. and Morgan, E. 1991. On-shore entrainment of circatidal rhythmicity in *Lipophyrrys pholis* by natural zeitgeber and the inhibitory effect of caging. *Mar. Behav. Physiol.*, 19: 63–73.

Page, T.L. 1982. Transplantation of the cockroach circadian pacemaker. *Science*, 216: 73–75.

Page, T.L. 1983a. Effects of optic-tract regeneration on internal coupling in the circadian system of the cockroach. *J. Comp. Physiol.*, 153: 353–363.

Page, T.L. 1983b. Regeneration of the optic tracts and the circadian pacemaker activity in the cockroach *Leucophaea maderae*. *J. Comp. Physiol.*, 152: 231–240.

Page, T.L. and Larimer, J.L. 1972. Entrainment of the circadian locomotor activity rhythm in the crayfish. *J. Comp. Physiol.*, 78: 107–120.

Page, T.L. and Larimer, J.L. 1975. Neural control of circadian rhythmicity in the crayfish. I. The locomotor rhythm. *J. Comp Physiol.*, 97: 59–80.

Page, T.L. and Nalovic, K.G. 1992. Properties of mutual coupling between the two circadian pacemakers in the eyes of the mollusc *Bulla gouldiana*. *J. Biol. Rhythms*, 7: 213–226.

Palmer, J.D. 1973. Tidal rhythms: the clock control of the rhythmic physiology of marine organisms. *Biol. Rev.*, 48: 377–418.

Palmer, J.D. 1974. *Biological Clocks in Marine Organisms*. Wiley-Interscience Publ., New York.

Palmer, J.D. 1988. Comparative studies of tidal rhythms. VI. Several clocks govern the activity of two species of fiddler crabs. *Mar. Behav. Physiol.*, 13: 231–243.

Palmer, J.D. 1989. Comparative studies of tidal rhythms. VII. The circalunidian locomotor rhythm of the brackish-water fiddler crab, *Uca minax*. *Mar. Behav. Physiol.*, 14: 129–143.

Palmer, J.D. 1990a. Comparative studies of tidal rhythms. IX. The modifying roles of deuterium oxide and azadiractin on circalunidian rhythms. *Mar. Behav. Physiol.*, 17: 167–175.

Palmer, J.D. 1990b. The rhythmic lives of crabs. *BioScience*, 40: 352–358.

Palmer, J.D. and Dowse, H.B. 1969. Preliminary findings on the effect of D_2O on the period of circadian activity rhythms. *Biol. Bull.*, 137: 388.

Palmer, J.D. and Williams, B.G. 1986. Comparative studies of tidal rhythms. II. The dual clock control of the locomotor rhythms of two decapod crustaceans. *Mar. Behav. Physiol.*, 12: 269–278.

Palmer, J.D. and Williams, B.G. 1987. Comparative studies of tidal rhythms. III. Spontaneous splitting of the peaks of crab locomotory rhythms. *Mar. Behav. Physiol.*, 13: 63–75.

Palmer, J.D. and Williams, B.G. 1993. Comparative studies of tidal rhythms. XII. Persistent photoaccumulation and locomotor rhythms in the crab, *Cyclograpsus lavauxi*. *Mar. Behav. Physiol.*, 22: 119–129.

Pardo, B.F. and Sáenz, E.M. 1988. Action of deuterium oxide upon the ERG circadian rhythm of the crayfish *Procambarus bouvieri*. *Comp. Biochem. Physiol.*, 90A: 435–440.

Petpiroon, S. and Morgan, E. 1983. Observations on the tidal activity rhythm of the periwinkle *Littorina nigrolineata*. *Mar. Behav. Physiol.*, 9: 171–192.

Ralph, M.R. and Menaker, M. 1988. A mutation of the circadian systems in golden hamsters. *Science*, 241: 1225–1227.

Ralph, M.R., Foster, R.G., Davis, F.C. and Menaker, M. 1990. Transplanted suprachiasmatic nucleus determines circadian period. *Science*, 247: 975–978.

Reddy, P., Zehring, W.A., Wheeler, D.A., Pirrotta, V., Hadfield, C., Hall, J.C. and Rosbash, M. 1983. Molecular analysis of the period locus in *Drosophila melanogaster* and identification of a transcript involved in biological rhythms. *Cell*, 38: 701–710.

Reid, D.G. and Naylor, E. 1989. Are there separate circatidal and circadian clocks in the shore crab *Carcinus maenas*? *Mar. Ecol. Prog. Ser.*, 52: 1–6.

Reid, D.G. and Naylor, E. 1993. Different free-running periods in split components of the circatidal rhythm in the shore crab *Carcinus maenas*. *Mar. Ecol. Prog. Ser.*, 102: 295–302.

Robertson, L.M. and Takahashi, J.S. 1988. Circadian clock in cell culture: 1. Oscillation of melatonin release from dissociated pineal cells in flow-through microcarrier culture. *J. Neurosci.*, 8: 12–21.

Roenneberg, T. and Morse, D. 1993. Two circadian oscillators in one cell. *Nature*, 362: 362–364.

Rodríguez-Sosa, L., Calderón, J., Hernández, J. and Aréchiga, H. 1990. The isolated eyestalk of the crayfish maintains a circadian rhythm of neurosecretion. *Soc. Neurosci. Abst.*, 16: 909.

Rusak, B. and Groos, G. 1982. Suprachiasmatic stimulation phase shifts rodent circadian rhythms. *Science*, 215: 1407–1409.

Sánchez, J. and Fuentes-Pardo, B. 1977. Circadian rhythm in the amplitude of the electroretinogram in the isolated eyestalk of the crayfish. *Comp. Biochem. Physiol.*, 56A: 601–609.

Sawaki, Y., Nihonmatsu, I. and Kawamura, H. 1984. Transplantation of the neonatal suprachiasmatic nuclei into rats with complete bilateral suprachiasmatic lesions. *Neurosci. Res.*, 1: 67–72.

Schweiger, H.-G., Hartwig, R. and Schweiger, M. 1986. Cellular aspects of circadian rhythms. *J. Cell. Sci.*, 4(suppl.): 181–200.

Shin, H.-S., Bargiello, T.A., Clark, F.T., Jackson, F.R. and Young, M. W., 1985. An unusual coding sequence from a *Drosophila* clock gene is conserved in vertebrate. *Nature*, 317: 445–448.

Silver, R., Lehman, M.N., Gibson, M., Gladstone, W.R. and Bittman, E.L. 1990. Dispersed cell suspensions of fetal SCN restore circadian rhythmicity in SCN-lesioned adult hamster. *Brain Res.*, 525: 45–58.

Siwicki, K.K., Eastman, C., Petersen, G., Rosbash, M. and Hall, J.C. 1988. Antibodies to the *period* gene product of *Drosophila* reveal diverse tissue distribution and changes in the visual system. *Neuron*, 1: 141–151.

Smietanko, A. and Englemann, W. 1989. Splitting of circadian rhythms of *Musca domestica* with azadirachtin. *J. Interdiscipl. Cycle Res.*, 20: 71–79.

Smith, G. and Naylor, E. 1972. The neurosecretory system of the eyestalk of *Carcinus maenas*. *J. Zool., Lond.*, 166: 313–321.

Stephen, F. and Zucker, I. 1972. Circadian rhythms in drinking behavior and locomotor activity are eliminated by hypothalamic lesions. *Proc. Natl. Acad. Sci.*, 69: 1583–1586.

Suter, R.B. and Rawson, K.S. 1968. Circadian activity rhythm of the deer mouse, *Peromyscus*: effect of deuterium oxide. *Science*, 160: 1011–1014.

Sweeney, B.M. 1974. The potassium content of *Gonyaulax polyedra* and phase changes in the circadian rhythm of stimulated bioluminescence by short exposures to ethanol and valinomycin. *Plant Physiol.*, 53: 337–347.

Takahashi, J.S., Hamm, H. and Menaker, M. 1980. Circadian rhythms of melatonin release from individual superfused chicken pineal glands *in vitro*. *Proc. Natl. Acad. Sci.*, 77: 2319–2322.

Takahashi, J.S. and Menaker, M. 1979. Physiology of avian circadian pacemakers. *Fed. Proc.*, 38: 2583–2588.

Takahashi, J.S. and Menaker, M. 1982. Role of the suprachiasmatic nuclei in the circadian system of the house sparrow, *Passer domesticus*. *J. Neurosci.*, 2: 815–828.

Takahashi, J.S. and Menaker, M. 1984. Multiple redundant circadian oscillators within the isolated avian pineal gland *J. Comp. Physiol.*, 154: 435–440.

Taylor, W., Gooch, V.D. and Hastings, J.W. 1979. Period shortening and phase shifting effects of ethanol on the *Gonyaulax* clock. *J. Exp. Physiol.*, 121: 355–358.

Taylor, W., Krasnow, R., Dunlap, J.C., Broda, H. and Hastings, J.W. 1982. Critical pulses of anisomycin drive the circadian oscillator in *Gonyaulax* towards its singularity. *J. Comp. Physiol.*, 148: 11–25.

Thompson, J.F. 1963. *The biological effects of deuterium*. Pergamon Press, New York.

Truman, J.W. 1972a. Circadian rhythms and physiology with special reference to neuroendocrine processes in insects. In: Bierhuizen, J.F. (Ed.), *Circadian Rhythmicity*, pp. 111–135. Centre for Agricultural Publishing and Documentation, Wageningen, The Netherlands.

Truman, J.W. 1972b. Physiology of insect rhythms. II. The silkmoth brain as the location of the biological clock controlling eclosion. *J. Comp. Physiol.*, 81: 99–114.

Truman, J.W. 1973. Physiology of insect ecdysis. II. The assay and occurrence of the eclosion hormone in the Chinese oak silkmoth, *Antheraea pernyi*. *Biol. Bull.*, 144: 200–211.

Truman, J.W. 1974. Physiology of insect rhythms. IV. Role of the brain in the regulation of the flight rhythm of the giant silkmoths. *J. Comp. Physiol.*, 95: 281–296.

Truman, J.W. and Riddleford, L.M. 1970. Neuroendocrine control of ecdysis in silkmoths. *Science*, 167: 1624–1626.

Vitaterna, M.H., King, D.P., Chang, A., Kornhauser, J.M., Lowrey, P.L., McDonald, J.D., Dove, W.F., Pinto, L.H., Turek, F.W. and Takahashi, J.S. 1994. Mutagenesis and mapping of a mouse gene, *Clock*, essential for circadian behavior. *Science*, 264: 719–725.

von der Heyde, F., Wilkens, A. and Rensing, L. 1992. The effects of temperature on the circadian rhythms of flashing and glow in *Gonyaulax polyedra*: are the two rhythms controlled by two oscillators? *J. Biol. Rhythms*, 7: 115–123.

Webb, H.M., Bennett, M.F. and Brown, F.A. 1954. Persistence of an endogenous diurnal rhythmicity in eyestalkless *Uca pugilator*. *Biol. Bull.*, 106: 371–377.

Welsh, J. 1930. Diurnal rhythm of the distal pigment cells in the eyes of certain crustaceans. *Proc. Nat. Acad. Sci.*, 16: 386–395.

Wever, R.A. 1979. *The Circadian Systems in Man: Results of Experiments Under Temporal Isolation*. Springer-Verlag, New York.

White, L., Ringo, J. and Dowse, H.B. 1992. A circadian clock of *Drosophila*: effects of deuterium oxide and mutations at the period locus. *Chronbiol. Int.*, 9: 250–259.

Wiedenmann, G. 1983. Splitting in the circadian activity rhythm: the expression of bilaterally paired oscillators. *J. Comp. Physiol.*, 150: 51–60.

Williams, B.G. 1991. Comparative studies of tidal rhythms. V. Individual variation in the rhythmic behaviour of *Carcinus maenas*. *Mar. Behav. Physiol.*, 19: 97–112.

Williams, B.G. 1995. Tidal biological clocks and their diel counterparts. In: Hartnoll, R.G. and Hawkins, S.I. (Eds), *Marine Biology — A Port Erin Perspective*, in press. Immel Publishing Co., London.

Williams, B.G. and Naylor, E. 1967. Spontaneously induced rhythm of tidal periodicity in laboratory-reared *Carcinus. J. Exp. Biol.*, 47: 229–234.

Williams, B.G., Palmer, J.D. and Hutchinson, D.N. 1993. Comparative studies of tidal rhythms. XIII. Is a clam clock similar to those of other intertidal animals? *Mar. Behav. Physiol.*, 24: 1–14.

Williams, J.A. 1985. Evaluation of optic tract section on the locomotor activity rhythm of the shore crab *Carcinus maenas. Comp. Biochem. Physiol.*, 82A: 447–453.

Williams, J.A., Pullin, R., Naylor, E., Smith, G. and Williams, B.G. 1979. The role of Hanström's organ in clock control in *Carcinus maenas*. In: Naylor, E. and Hartnoll, R.G. (Eds), *Cyclic Phenomena in Marine Plants and Animals*, pp. 459–466. Pergamon Press, Oxford.

Zerr, D.M., Hall, J.C., Rosbash, M. and Siwicki, K.K. 1990. Circadian fluctuation of period protein immunoreactivity in the CNS and visual system of *Drosophila. J. Neurosci.*, 10: 2749–2762.

Zimmerman, N.H. and Menaker, M. 1975. Neural connections of sparrow pineal: role in circadian control of activity. *Science*, 190: 477–479.

Glossary

Amplitude: A measurement of the height of the peaks relative to the troughs of a cycle. The measurement is usually made from the cycle mean to an extreme.

Biological clock: A pacemaking mechanism that produces those organismic rhythms that will persist in constant conditions.

Chronobiologist: A life scientist who specializes in the study of biological rhythms. Those chronobiologists who work with tide-related organismic rhythms can be easily identified by their haggard look due to loss of sleep, by expressions of dejection from seldom being able to repeat an experimental result, and by their obvious enjoyment of martinis.

Chronobiology: The subject of biological rhythms.

Circadian rhythm: A contraction of *circa*, which means "about": and *dies*, which means "a day". Such a rhythm is a basic solar-day one whose period has either slightly lengthened or shortened in constant conditions. These rhythms can be entrained by light/dark cycles ranging from about 22 h to 26 h in length.

Circalunidian-clock hypothesis: The postulate that tide-associated organismic rhythms are controlled by two clocks, each running at a basic period of 24.8 h. The clocks are coupled to one another 180° out of phase. Each clock controls one of the two tidal peaks displayed by animals each lunar day. In constant conditions the periods of their rhythms become "circa." Because of the mutual coupling, the combined output of these two clocks is indistinguishable from the output of a circatidal clock. But their existence can become manifest in the laboratory when the coupling between them ruptures; then the two peaks can be seen to have become independent of one another. An example is seen in Fig. 3-8 where the peaks have adopted different period lengths.

Circalunidian rhythm: A basic lunar-day rhythm that has become slightly longer or shorter than the length of a lunar day in constant conditions. The "circa" period of circadian and circalunidian rhythms overlaps greatly, but the two can be distinguished by the fact that circalunidian rhythms cannot be entrained by light/dark cycles, and circadian rhythms cannot be entrained by physical and chemical parameters such as hydrostatic pressure and salinity cycles.

Circatidal-clock hypothesis: A noetic postulate that tide-associated organismic rhythms are controlled by a single clock with a basic period of 12.4 h — the average interval of most tides. In constant conditions the period of the rhythm becomes "circa."

Circatidal rhythm: An organismic rhythm that has become slightly longer or shorter than 12.4 h in constant conditions.

Constant conditions: A laboratory setting in which at least the levels of illumination and temperature do not vary. Needless to say, there are no tides in the lab.

Cycle: A sequence of events that repeats itself through time in the same order and at the same interval. In this book it is synonymous with rhythm.

Discombobulated: A common mental state seen in marine chronobiologists after a day of trying to decipher typical tidal-rhythm data.

Entraining agent: an environmental cyclic event, such as day/night alternation, or the flood and ebb of the tides, that locks the phase of an organismic rhythm to the agent's phase and periodicity.

Form estimate: The averaging of several consecutive cycles together. It is a technique commonly used to summarize and envisage the general shape of a rhythm (Fig. 2-7b).

Fortnight: In the United Kingdom, two weeks; here, an interval of 14.75 days.

Frequency: The number of cycles per some unit of time; the inverse of period.

Intertidal zone: The littoral region that is above the low-water mark and below the high-water mark. The subtidal region is the tideless zone just below the low-water mark. The supratidal zone is that area of the shore that is just out of reach of the high tides.

Lunar day: The 24.8-h average interval between, say, consecutive moonrises; one rotation of the earth relative to the moon (Fig. 1-2).

Lunar-day rhythm: An organismic rhythm with a period of 24.8 h.

Meta-oscillator: A living clock consisting of an ensemble of short-period, intracellular oscillators coupled together in such a way as to produce overt rhythms of circadian and circalunidian periods.

Neap tides: Those tides occurring twice each month — during the quadrates — when the magnitude (= range) of tidal exchange is the smallest. Saying it another way, the times of the month when the water at high tides does not rise very far up on the shore and does not retreat very far at low tide (Fig. 1-5).

Period: The time interval of one complete cycle.

Persistent rhythm: Any organismic rhythm that will continue to be expressed in the laboratory in constant conditions.

Phase: Some arbitrarily chosen fraction (such as the peak or trough) of a cycle.

Phase-response curve: A graph of the direction and amount of phase change produced in a rhythm plotted against the time at which the rhythm was subjected to a phase-change-producing stimulus (Fig. 4-14). Usually initialized as PRC.

Quadrature: A 90° angle of alignment between the sun, moon and earth (Fig. 1-4b). In this configuration the tractive forces of the sun and moon work against each other, resulting in the creation of the low-amplitude neap tides.

Rhythm: Same as cycle.

Semidiurnal inequality: A condition in which the two tidal inundations during each lunar day are of unequal amplitude (Figs 1-3 and 1-5). In some cases organisms subjected to these tides show the same inequality in peak height of their persistent rhythms in the laboratory (Fig. 4-15).

Solar day: The 24-hour interval between, say, consecutive sunrises. One rotation of the earth relative to the sun.

Solar-day rhythm: An organismic rhythm with a period length of 24 h.

Spring tides: Those tides occurring twice each month — during the syzygies — when the range of tidal exchange is the greatest. These are the times of the highest, high tides, alternating with the lowest, low tides (Fig. 1-5).

Synodic month: The 29.5-day interval between, say, consecutive new-moon phases.

Syzygy: A wonderful word to know when playing Scrabble. An alignment of the sun, moon and earth in a straight line (Fig. 1-4a). In this configuration the tractive forces of the sun and moon are additive, creating the spring tides.

Transients: An interval of changing period length seen in some organisms between the time a phase-resetting stimulus is applied and a new steady-state period is eventually assumed.

Ultradian rhythm: A rhythm with a period length shorter than the lower range of circadian or circalunidian rhythms, and longer than most metabolic oscillatory processes. In this book they arbitrarily include periods between 1 h and 10 h.

Zeitgeber: Literally, a "time giver." Any environmental stimulus that will entrain or rephase a biological rhythm.

Author Index

Abelló, P. 96, 105, 132, 134
Ables, J. 18, 24
Al-Adhub, A.H.Y. 105, 132
Aldrich, J.C. 31, 83
Aréchiga, A. 185–8, 196, 199, 201
Arudpragasam, K.D. 44, 83
Atkinson, R.J. 44, 83, 92–4, 132–4

Bargiello, T. A. 171, 196, 201
Barnwell, F.H. 35, 41, 83, 193, 196
Barrera-Mera, B. 189, 196
Bartels, J. 16, 30
Bayne, B. L. 63, 87
Beentjes, M. P. 63–4, 83
Beer, C.G. 56, 83
Bennett, M.F. 15–17, 29–30, 32,
    63, 83, 184, 196, 202
Bennitt, R. 187, 196
Benzer, S. 170, 198–9
Bergin, M.E. 148, 157
Binkley, S.A. 5, 13
Bittman, E.L. 173, 198, 201
Blackman, R.B. 18, 29
Bliss, D.E. 145, 157
Block, G.D. 175–6, 189, 196–9
Bolt, S.R. 93–7, 100–4, 132, 134

Boulos, Z. 5, 13
Bowers, R. 126–8, 133
Brown, F.A. 15–17, 29–30, 32, 63,
    83, 141–2, 157, 184, 196, 202
Brown, R.A. 29, 83
Bruce, V.G. 165–6, 196
Buggy, J. 173, 196
Bünning, E. 5, 13

Caspers, H. 136, 158
Chapman, S. 16, 30
Chatfield, C. 18, 30
Chatterton, T.D. 56, 88
Christy, J.H. 148, 158
Citri, Y. 171, 196
Cole, L.C. 15, 30
Cordiner, S. 131, 133
Crisp, D.J. 69, 87
Cymborowski, B. 178, 196

Daan, S. 35, 40, 86
Davids, E. 63, 83
Davies, D.A. 103, 134
DeCoursey, P.J. 5, 13, 150, 156,
    158, 173, 196

Defant, A. 12, 13
Douglass, J.K. 105, 133
Dowse, H.B. 16–18, 25, 30, 43, 58–61, 83, 166, 170–2, 197, 202
Dunlap, J.C. 163, 171, 197, 202

Edmunds, L.N. 5, 13, 162, 197
Edwards, D.H. 188, 197
Edwards, G.A. 184, 197
Eichler, V.B. 172, 199
Eik-Nes, K.B. 73, 87
Englemann, W. 167–8, 198, 201
Enright, J.T. 15, 17, 30, 63, 83, 109–13, 120, 132–3, 164, 166, 169, 193, 197
Eskin, A. 177, 197
Evans, J.W. 69, 83
Evans, W.G. 81, 84
Ewer, J. 170, 197

Feldman, J.F. 162, 171, 197
Field, C.B. 195, 198
Fielder, D.R. 56, 84
Fincham, A.A. 120, 133
Fingerman, M. 30, 50, 84, 184, 197
Fish, J.D. 63, 84
Fish, S. 63, 84
Fisher, A. 136, 159
Fleissner, G. 179, 181–2, 190–2, 198
Forward, R.B. 105, 133
Franke, H.-D. 138–40, 158, 197
Fuentes-Pardo, B. 188–9, 197, 201
Fuller, C.A. 106, 133

Garcia, E.S. 167, 198
Gaston, S. 167, 173, 198
Gibson, R.N. 72–3, 84–5, 122–31, 133–4, 193, 199
Gifford, C.A. 148, 158
Goodenough, J.E. 142, 153, 158, 160, 167, 198
Goodwin, B. 42, 84
Graham, J.M. 126–8, 133
Green, J.M. 124, 133, 193, 198
Groos, G. 173, 201

Gwinner, E. 5, 13, 40, 84

Hall, J.C. 170–1, 197–201
Handler, A.M. 178, 198
Han, S.-Z. 167–8, 198
Happey-Wood, C. 81, 84
Hardin, P.E. 171, 198
Harris, G.J. 17, 30, 62, 84, 115–20, 133, 165, 169–70, 191, 198
Hastings, J.W. 5, 13, 30, 157, 163–4, 202
Hastings, M.H. 17, 63, 84, 114, 133, 141, 150–2, 159, 199
Hauenschild, C. 136, 159
Hayes, C.J. 166, 198
Heimbach, F. 154, 156–9
Hennessey, T.L. 195, 198
Herdman, E.C. 75, 84
Hidaka, T. 143, 160
Hoffman, K. 40, 84
Hofmann, D. 136, 159
Holmström, W.E. 62, 84, 114–17, 133
Holst, E.V. 54, 85
Hoyle, M. 171, 197
Huberman, A. 186, 196
Hutchinson, D.N. 65–8, 85, 88, 120–1, 134, 203

Jacklet, J.W. 175–6, 198
Jones, D.A. 63, 85, 112–14, 133
Jones, M.B. 56, 84
Jorgensen, C.B. 63, 85

Kalmus, H. 187, 198
Keeble, F. 140, 159
Keeton, W.T. 143, 159
Keller, S. 164, 198
Kenney, B.E. 105, 133
Kerfoot, W.B. 152, 159
Klapow, L.A. 109–12, 133, 198
Koehler, W.K. 179, 181–2, 198
Konopka, R.J. 170, 178, 198
Korringa, P. 135–6, 152, 159
Kupfermann, I. 175, 198

Larimer, J.L. 187, 189, 200

Larkin, T. 143, 159
Lehman, M.N. 173, 198, 201
Lickey, M. 176, 199
Liu, W. 170, 199
Li-Weber, M. 171, 199
Lohman, K.J. 143, 159
Lutz, E. 69, 85

Mangerich, S. 188, 199
McDaniel, M. 166, 199
McMahon, D.G. 176, 196
Menaker, M. 171, 173–4, 198, 200–3
Mercer, D.M. 18, 30
Mergenhagen, D. 183, 195, 199
Michels, S. 176, 199
Miller, C.D. 138, 159
Moore, R.Y. 172, 199
Moore-Ede, M.C. 5, 13, 106, 133
Moran, V.C. 152–3, 160
Morgan, E. 17, 30, 62, 73, 84–5, 114–20, 123–34, 165, 169–70, 191, 193, 198–200
Morris, N.M. 73–5, 86
Morse, D. 195, 201
Morton, B.S. 63, 85

Nagabhushanam, R. 50, 84
Nalovic, K.G. 195, 200
Naylor, E. 17, 21, 24, 30–1, 43–50, 56, 63, 83, 85, 87–9, 93–105, 108–9, 112–14, 134, 150–2, 156, 158–60, 184–6, 189, 192, 196, 199–201, 203
Neumann, D. 153, 159
Nishiitsutsuji-Uno, J. 178, 199
Noguerón, I. 188, 199
Northcott, S.J. 73, 85, 123–31, 134, 193, 199

Oehmke, M.G. 152, 159

Page, T.L. 54, 85, 178–180, 187, 189, 195, 199–200.
Palmer, J.D. 5–6, 11–13, 18, 21, 23, 26–7, 30, 34–43, 50–6, 65–9, 73–80, 82–3, 85–8, 120–1, 134, 136, 140–1, 148, 153, 157, 160, 166–9, 186–7, 193, 197–8, 200, 203
Palmer, R.E. 63, 86
Pardo, B.F. 188, 200
Park, O. 82, 86
Park, Y.H. 142, 157
Parsons, A.J. 44, 83, 93–4, 132
Pavlidis, T. 42, 86
Pearse, J.E. 156, 160
Pen, F. 138, 159
Pengelley, E.T. 5, 13
Petpiroon, S. 193, 200
Philpott, L. 50, 84
Pittendrigh, C.S. 35, 40, 86, 165–6, 178, 196, 199
Pullin, R.S. 185, 196, 203

Ralph, C.L. 63, 83
Ralph, M.R. 173, 200
Rao, K.P. 63, 87
Rawson, K.S. 166, 201
Reaka, M.L. 156, 160
Reddy, P. 170, 200
Redfield, A.C. 12, 13
Reichle, D.E. 82, 87
Reid, D.G. 63, 87, 92, 96–105, 132, 134, 150–2, 160, 192, 200–1
Rensing, L. 195, 202
Resko, J.A. 73, 87
Rhoads, D.C. 69, 85
Richardson, C.A. 69, 87
Riddleford, L.M. 177, 202
Ringo, J.M. 16–18, 25, 30, 43, 83, 166, 170–2, 197, 202
Robertson, L.M. 174, 201
Robinson, E. 17, 31, 74
Rodriguez-Sosa, L. 188, 201
Roenneberg, T. 195, 201
Rosbash, M. 170, 171, 196–201, 203
Round, F.E. 76–80, 86–7
Runham, N.W. 69, 87
Rusak, B. 5, 13, 173, 201

Saigusa, M. 143–9, 160
Sandeen, M.I. 16, 30
Sanders, D.S. 139, 160

Sawaki, Y. 173, 201
Sánchez, J. 188, 201
Schweiger, H.-G. 171, 183, 195, 199, 201
Schweiger, M. 183, 201
Seiple, W. 51, 87
Shin, H.-S. 171, 201
Shriner, J. 63, 83
Silver, R. 173, 201
Simon, R.B. 82, 87
Siwicki, K.K. 171, 201, 203
Smetzer, B. 137, 160
Smietanko, A. 167–8, 201
Smith, G. 184, 199, 201, 203
Southern, A.L. 73, 87
Sprague, P.C. 145, 157
Stephan, F. 172, 201
Sulzman, F.M. 106, 133, 166, 199
Suter, R.B. 166, 201
Sweeney, B.M. 5, 13, 164, 169, 201

Takahashi, J.S. 171, 174–5, 197, 201–2
Taylor, A.C. 92–3, 134
Taylor, W.R. 78, 87, 163–4, 202
Thompson, J.F. 166, 202
Thompson, R.J. 63, 87
Truman, J.W. 177, 202
Tukey, J.W. 18, 29
Turek, F.W. 40, 87, 171, 202

Udrey, J.R. 73–5, 86

Underwood, H. 40, 87
Vitaterna, M.H. 171, 202
von der Heyde, F. 195, 202

Warman, C.G. 43, 45, 62, 87, 96, 134
Webb, H.M. 16, 30–2, 41, 63, 83, 87, 184, 196, 202
Wells, J.W. 68–9, 87
Welsh, J. 188, 202
Wever, R.A. 195, 202
Wheeler, D.E. 148, 160
White, L. 166, 202
Whittaker, E. 17, 31
Wiedenmann, G. 194, 202
Williams, B.G. 18, 23, 30, 42, 45–50, 54, 56–68, 83, 86–9, 92, 94, 107–9, 120–3, 134, 184–6, 189, 192, 193, 199–200, 202
Williams, J.A. 17, 24, 30, 45, 88, 92, 134, 150, 160, 185, 196, 203
Willows, A.O. 143, 159
Wylie, F.E. 12, 13

Young, M.W. 171, 196, 201
Youthed, G.J. 152–3, 160

Zerr, D.M. 171, 203
Zimmerman, N.H. 173, 203
Zucker, I. 172, 201

Subject Index

Acetabularia
  multiple clocks 3, 183
  rhythmic cell fractions 183
Acheta domesticus, brain clock 178
Actograph, 33, 127
Anisomycin 164
Ant lion, see *Myrmeleon obscurus*
Antheraea pernyi, brain clock 177
Aphis mellifica carnica, fortnightly
    and monthly rhythms 152
Aplysia californica
  activity rhythm 174, 176
  neuronal clock in isolation 175–6
  neuronal-firing rhythm 174–6
  ultradian rhythm 176
Array analysis, method 26–9
Aschoff's Rule 4
Austrovenus stutchburyi
  entrainment with inundation
      cycles 120–3
  rephasing with temperature 120–1
  shell-gaping rhythm 63–9
Autocorrelation analysis 18–26
Azadirachtin 167–9

Basket cockle, see *Clinocardium*

Bathyporeia pelagica, swimming
    rhythm 120
Biological clock
  accuracy 12, 14
  adaptive nature, 12, 193
  locations 3, 104
  properties 4, 192–3
Blaps gigas
  light-sensitivity rhythm 179, 181
  spontaneous change in period
      179, 181
BMDP-IT analysis 18–26
Bulla gouldiana, single-cell clock
    176

Carcinus maenas
  blood-sugar rhythm 44–5, 92
  effect of eyestalk removal 93,
      184–6
  entrainment by inundation cycles
      89–90
  entrainment by pressure cycles
      90–2, 102–3
  entrainment by salinity cycles
      92–104

Carcinus maenas—(cont.)
  entrainment by temperature cycles
    90, 100–4
  initiation of rhythm with
    temperature pulse, 45–7
  oxygen-consumption rhythm 44
  peak splitting 98–9, 100
  solar-day-activity rhythm 43–6
  temperature compensation of
    period 47
  tide-associated activity 43–50
Carcinus mediterraneus
  activity rhythm 45
  non-entrainment by salinity cycles
    96
Cardisoma guanhumi, monthly
  larval-release rhythm 148
Cave cricket, see *Ceuthophilus*
Cerastoderma edule
  fortnightly rhythm 69
  shell-gaping rhythm 69
Ceuthophilus maculatos, activity
  rhythm 81–3
Circa nature of rhythms,
  definition 4
Circalunidian-clock hypothesis 41–2,
  56, 59, 68, 73, 80, 105, 112,
  121–3, 124, 128, 131, 169,
  192–4
Circatidal-clock hypothesis 192
Clinocardium nuttalli, shell-gaping
  rhythm 68
Clockle, see *Austrovenus*
Clocks, persistent funtioning in
  isolation 174–9, 183
Cloudy bubble snail, see *Bulla*
Clunio marinus
  fortnightly reproductive rhythm
    153–6
  rhythm set by agitation cycles
    154–6
  rhythm set by temperature cycles
    156–7
Commuter diatom, see *Hantzschia*
Compact plot, method of
  construction 28
Convoluta roscofensis
  fortnightly gamete-release rhythm
    140–1
  vertical migration rhythm
    140–1

Corophium volutator
  activity rhythm 59
  entrainment by inundation cycles
    114–15
  entrainment by temperature cycles
    116–17
  ethanol phase-response curve 164
  ineffectiveness of cycloheximide
    169–70
  oxygen-consumption rhythm 62
  salinity phase-response curve
    117–20
  temperature phase-response curve
    115–16
  using temperature to locate the
    clock 189–91
  valinomycin phase-reesponse
    curve 164, 169
Coupler, clock, 4, 35, 41, 64, 195
Crangon crangon, activity rhythm
  105
Cranny crab, see *Cyclograpsus*
Crassostrea virginica, shell-gaping
  rhythm 63
Crayfish clocks, 187, 189
  optic-pigmentation-migration
    rhythms 187–8
Cyclograpsus lavauxi
  rhythm splitting 54
  solar-day activity component 54
  tide-associated activity 54, 57
Cycloheximide 169–70

Deuterium oxide 165–7
Drosophila melanogaster, brain clock
  178
Dugesia dorotocephala, monthly
  orientation rhythm 141–2

Entrainment
  by agitation cycles 109–13,
    112–14, 130–2, 144, 150–2,
    154–6
  by inundation cycles 89–90,
    114–15, 120–6
  by pressure cycles 90–2, 102–3,
    105, 112–14
  by salinity cycles 92–104
  by temperature cycles 90, 100–4,
    107–9, 116–17, 154–6

Eunice viridis, gamete-release rhythm
136–8
Eurydice pulchra
activity 62
entrainment by agitation cycles
112–14
entrainment by pressure cycles
112
fortnightly activity rhythm 150–2
oxygen-consumption rhythm 62
phasing fortnightly rhythm with
agitation 151–2
swimming rhythm 112–14
Excirolana chiltoni
entrainment by agitation cycles
109–13
ineffectiveness of cycloheximide
169
period lengthening with alcohol
165
period lengthening with
deuterium 166
phase-response curve 111
swimming rhythm 109–13
Exogenous/endogenous clock
controversy 35–6

Fiddler crab, see *Uca*
Form-estimate, method of
construction 16, 27, 38, 49
Fortnightly rhythms 63, 69, 140–1,
143–52
phase set by light pulses 144,
147, 153–5

Geochronometers 68–9
Gobius paganellus
activity rhythm 73
entrainment by pressure cycles
129
Goby, see *Gobius*
Gonyaulax polyedra
bioluminescent-glow rhythm 163
period lengthening with
deuterium 166
period shortening with alcohol
164
phase change with anisomycin
164

Hanström's organ 184
Hantzschia virgata
tidal rhythm in phototaxis 78
vertical-migration rhythm 75–9
Heavy water 165–7
Helice crassa, activity rhythm 56–9
Hemigrapsus edwardsi
activity rhythm 107–9
phase-response curve 108
Humans, daily rhythm in copulation
73–5
Hyalophore ceropia, brain clock
176–7

Innate nature of tidal rhythms 45–7

Leucophaea maderae
locomotor rhythm 178–80
optic-lobe clock 178–9
Liocarcinus holsatus, activity rhythm
105
entrainment by pressure cycles
105
Lipophrys pholis
activity rhythm 72
entrainment by agitation cycles
130–2
entrainment by inundation cycles
122–6
entrainment by pressure cycles
125–9
pressure phase-response curve
130–1
Lunar day, generation of 7
Lunar-day and solar-day clocks,
one-and-the-same 194–6

Macrophthalmus hirtipes, activity
rhythm 27, 56–61
Mercenaria mercinaria, shell-gaping
rhythm 63
MESA analysis 18–26
Meta-oscillator 172, 195
Monthly rhythms 136, 138–9,
141–3, 148, 152–4
phased by light pulses 139

Myrmeleon obscurus, lunar-day and monthly dwelling-shape rhythm 152–3

Mytilus californianus, shell-gaping rhythm 63

Mytilus edulis, shell-gaping rhythm 63

Natural period, inadequacy of term 35–6

Neurodepressing hormone (NDH) 184–6

Non-adaptive properties of solar- and lunar-day clocks 194

Oligocottus maculosus, effect of translocation on rhythm 193

Ostrea edulis, crystalline-style-volume rhythm 63

Paleontological clocks 68–9

Penultimate-hour crab, see *Sesarma*

Period, spontaneous change 28, 36, 39–41, 58–9, 66–8, 179, 181, 192

Periodogram analysis 17–25

Periplaneta americana, optic-lobe clock 178

Phase-response curve, calculation thereof 106–7

Phase-response curves 106, 108, 115–16, 117–20, 130

Pigeon, monthly homing orientation rhythm 143

Pigment-dispursing hormone (PDH) 184, 188

Pineal gland as a clock 173–4

Pleurosigma angulatum, vertical-migration rhythm 81

Pliant-pendulum crab, see *Helice*

Rhithropanopeus harrissii entrainment by salinity cycles 105

larval-release rhythm 105

Rocky shore crab, see *Hemigrapsus*

Samoan Palolo worm, see *Eunice*

Schizo crab, see *Macrophthalmus*

Sea hare, see *Aplysia*

Sesarma haematocheir, fortnightly larval-release rhythm 143–8

Sesarma reticulatum activity rhythm 50–5

search for the clock 186–7

solar-day activity component 50–4

splitting 51, 53

Shanny, see *Lipophrys*

Shore crab, see *Carcinus*

Solar-day and lunar-day clocks, one-and-the-same 194–6

Solar-day clocks, several periods in same individual 195

Sphecodogstra texana, monthly activity rhythm 152

Splitting, rhythm peak 36–40, 53, 54, 58, 98–9, 168–9, 192

Suprachiasmatic nuclei as a clock 172–3

Synchelidium sp., lack of entrainment by pressure cycles 120

Talitrus saltator, fortnightly activity rhythm 150

Temperature compensation of the period 4, 47

Thalassotrechus barbarae, activity rhythm 81

Tides, *circa* nature 12, 194

Tides, generation of various forms 7–10

Translocation, effect on phase 124, 193

Tritonia diomedea, monthly orientation rhythm 143

Typosyllis prolifera endocrine clocks? 190–2

gamete-release rhythm 138–40

U-oscillators 172, 195

Uca minax activity rhythm 36, 39, 40

fortnightly larval-release rhythm 149

Uca pugilator
activity rhythm 40
eyestalk clock? 183–4
fortnightly larval-release rhythm 149
oxygen-consumption rhythm 32
period lengthening with deuterium 166–7

Uca pugnax
activity rhythm 5, 6, 15, 17, 34–7, 43
effects of azadiractin 168
fortnightly larval-release rhythm 149
lack of effect of ethanol 168

oxygen-consumption rhythm 32
period lengthing with deuterium 166
ultradian rhythm 43
Ultradian rhythms 42–3, 58–62, 171–2, 194–5

Valinomycin 164, 169
Vertical-migration rhythms 75–80, 140–1, 176

X-organ/sinus-gland complex 184